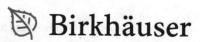

Frontiers in Mathematics

Advisory Board

Leonid Bunimovich (Georgia Institute of Technology, Atlanta)
William Y. C. Chen (Nankai University, Tianjin)
Benoît Perthame (Sorbonne Université, Paris)
Laurent Saloff-Coste (Cornell University, Ithaca)
Igor Shparlinski (The University of New South Wales, Sydney)
Wolfgang Sprößig (TU Bergakademie Freiberg)
Cédric Villani (Institut Henri Poincaré, Paris)

More information about this series at http://www.springer.com/series/5388

José Antonio Ezquerro Fernández •
Miguel Ángel Hernández Verón

Mild Differentiability Conditions for Newton's Method in Banach Spaces

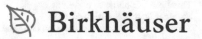

José Antonio Ezquerro Fernández
Department of Mathematics and Computation
University of La Rioja
Logroño, Spain

Miguel Ángel Hernández Verón
Department of Mathematics and Computation
University of La Rioja
Logroño, Spain

ISSN 1660-8046   ISSN 1660-8054   (electronic)
Frontiers in Mathematics
ISBN 978-3-030-48701-0   ISBN 978-3-030-48702-7   (eBook)
https://doi.org/10.1007/978-3-030-48702-7

Mathematics Subject Classification: 47H99, 65J15, 65H10, 45G10, 34B15, 35J60

This book is published under the imprint Birkhäuser, www.birkhauser-science.com, by the registered company Springer Nature Switzerland AG.
The registered company address is: Gewerbestrasse 11, 6330 Cham, Switzerland

To my daughter María,
the pillar of my life. JAE.

To my dear and always remembered parents
Carmen and Miguel. MAHV.

Preface

Many scientific and engineering problems can be written as a nonlinear equation $F(x) = 0$, where F is a nonlinear operator defined on a nonempty open convex subset Ω of a Banach space X with values in a Banach space Y. The solutions of this equation can rarely be found in closed form, so that we usually look for numerical approximations of these solutions. As a consequence, the methods for solving the previous equation are usually iterative. So, starting from one initial approximation of a solution x^* of the equation $F(x) = 0$, a sequence $\{x_n\}$ of approximations is constructed such that the sequence $\{\|x_n - x^*\|\}$ is decreasing and a better approximation to the solution x^* is then obtained at every step. Obviously, one is interested in $\lim_n x_n = x^*$.

We can then obtain a sequence of approximations $\{x_n\}$ in different ways, depending on the iterative methods that are applied. Among these, the best known and most used is Newton's method, whose algorithm is

$$x_{n+1} = x_n - [F'(x_n)]^{-1} F(x_n), \quad n \geq 0, \quad \text{with } x_0 \text{ given.}$$

It is well known that three types of studies can be carried out when we are interested in proving the convergence of Newton's sequence $\{x_n\}$ to the solution x^*: local, semilocal and global. First, the local study of the convergence is based on imposing conditions on the solution x^*, based on certain conditions on the operator F, and provides the so-called ball of convergence [25] of the sequence $\{x_n\}$, which shows the accessibility to x^* from the initial approximation x_0 belonging to the ball. Second, the semilocal study of the convergence is based on imposing conditions on the initial approximation x_0, based on certain conditions on the operator F, and provides the so-called domain of parameters [32] corresponding to the conditions required to the initial approximation that guarantee the convergence of the sequence $\{x_n\}$ to the solution x^*. Third, the global study of the convergence guarantees, based on certain conditions on the operator F, the convergence of the sequence $\{x_n\}$ to the solution x^* in a domain and independently of the initial approximation x_0. The three approaches involve conditions on the operator F. However, requirement of conditions on the solution, on the initial approximation, or on none of these, determines the different types of studies.

The local study of the convergence has the disadvantage of being able to guarantee that the solution, that is unknown, can satisfy certain conditions. In general, the global study of the convergence is very specific as regards the type of underlying operators, as a consequence of absence of conditions on the initial approximation and on the solution. There is a plethora of studies devoted to the weakening and/or extension of the hypotheses made on the underlying operators. In this monograph, we focus on the analysis of the semilocal convergence of Newton's method.

Three types of conditions are required to obtain semilocal convergence results for Newton's method: conditions on the starting point x_0, conditions on the underlying operator F and conditions that the two proceeding types of conditions. An important feature of the semilocal convergence results is that conclusions about the existence and uniqueness of solution of the equation to be solved can be drawn based on the theoretical result and the initial approximation. This fact makes the choice of the starting points for Newton's method a basic aspect in semilocal convergence studies.

The generalization of Newton's method to Banach spaces is due to the Russian mathematician L. V. Kantorovich, who was the first researcher to study the semilocal convergence of Newton's method in Banach spaces by publishing some several in the mid-twenty century, [49–57], and giving an influential result, known as *the Newton–Kantorovich theorem*. This gave rise to what is now known as *Kantorovich's theory*.

In our monograph *Newton's Method: An Updated Approach of Kantorovich's Theory*, [37], we analyse Kantorovich's theory based on the well-known *majorant principle* developed by Kantorovich, which in turn is based on the concept of *majorizing sequence*. There we present an adapted approach of this theory that includes old results, refines old results, proves the most relevant results and gives alternative approaches that lead to new sufficient semilocal convergence criteria for Newton's method. As we can see in that monograph, if we pay attention to the type of conditions required for the operator F to guarantee the semilocal convergence of Newton's method, there are conditions on F', as well as conditions on F'' or even conditions on successive derivatives of F.

However, if we look at the algorithm of Newton's method, we see that only the first derivative F' of the operator F is involved, so one should try to prove the semilocal convergence of the method by imposing conditions only to F'. If we proceed in this way, then the technique based on majorizing sequences of Kantorovich cannot be used to prove the semilocal convergence of Newton's method in all the situations that can be encountered. So, in the present monograph, we focus our attention on the analysis of the semilocal convergenece of Newton's method under mild differentiability conditions on F' and use a technique based on recurrence relations which is different from that based on majorizing sequences and which was introduced and developed by us over the years. As a consequence, we improve the domains of parameters associated with the Newton–Kantorovich theorem and other existing semilocal convergence results for Newton's method which are obtained under mild differentiability conditions on F'. In addition, center conditions on the operator F' play an important role in the study of the semilocal convergence of Newton's method, since we can improve the domain of starting points when the technique of recurrence relations is used to prove semilocal convergence.

This monograph is addressed to researchers interested in the theory of Newton's method in Banach spaces. Each chapter contains several theoretical results and interesting applications to nonlinear integral and differential equations.

We begin the monograph with a quick overview of Newton's method in Chap. 1, presenting a brief history of the method that ends with Kantorovich's theory, where the Newton–Kantorovich theorem is remembered and illustrated with an application to a Hammerstein integral equation. Then, we define what we mean by accessibility of an iterative method, defining the three ways in which it can be seen: basin of attraction, region of accessibility and domain of parameters. We finish the chapter by introducing mild differentiability conditions on F' as generalizations of Kantorovich's condition on F', along with a technique based on recurrence relations, that is used throughout the monograph as an alternative to the majorant principle of Kantorovich, to prove the semilocal convergence of Newton's method.

In Chap. 2, we develop the technique based on recurrence relations to prove the semilocal convergence of Newton's method when F' is Lipschitz continuous in the domain of definition of the operator F and conclude with an application to a Chandrasekhar integral equation.

The first generalization of the condition that F' is Lipschitz continuous is presented in Chap. 3, where we require that F' is Hölder continuous in the domain of definition of the operator F. We do an analysis similar to that given in the previous chapter for the Lipschitz case, complete this analysis with a comparative study involving semilocal convergence results given by other authors, and finish with an application that highlights how the theoretical power of Newton's method is used to draw conclusions about the existence and uniqueness of a solution and about the region in which it is located.

Chapter 4 contains a variant of the Hölder continuity condition on F' discussed in Chap. 3 that includes the Lipschitz and Hölder cases as special ones and leads to a modification of the domain of starting points, obtained previously, coming to a greater applicability of the method.

Chapter 5 introduces what we call the ω-Lipschitz continuous operators and analyzes the semilocal convergence of Newton's method when F' is ω-Lipschitz continuous in the domain of definition of the operator F. This condition includes, besides the Lipschitz and Hölder cases, the case in which F' is a combination of operators such that F' is Lipschitz or Hölder continuous, which often occurs for some nonlinear integral equations of mixed Hammerstein type.

We show in Chaps. 6 and 7 the important role played by the previous conditions when they are centered at the starting point x_0 of Newton's method, which leads to an improvement of the domain of starting points. We complete this analysis by comparing our results with results by other authors and highlight the importance of the domain of parameters associated with a semilocal convergence result. We illustrate the conclusions given applications to conservative problems and mildly nonlinear elliptic equations.

The fact that the conditions imposed to the starting point and to the operator F are independent enables us to can choose the initial approximation inside a domain of starting points depending on the conditions that the two types of hypotheses. In another case, if the two types of conditions are connected, the domain of starting points can be significantly reduced and this is a problem. In Chap. 8, we try to solve this problem by introducing an auxiliary point, different from the starting point, which allows us to eliminate the connection between the conditions required for the starting points and those required for the operator F, and thus recover the domain of starting points.

Applications to nonlinear integral and differential equations are included to motivate the ideas presented and illustrate the results given. In particular, we consider Hammerstein integral equations, conservative problems and elliptic equations, which are solved by discretization.

We have developed all the proofs presented in the monograph for a better understanding of the ideas presented, so that the reading of the monograph follows without difficulty. All the ideas presented in the monograph have been developed by us over the recent years and references to our work as well as to works of other researchers are provided in the bibliography.

Finally, throughout the monograph, we pursue to extend the application of Newton's method from the modification of the domain of starting points. To end, we impose various conditions on the operator involved and used a technique based on recurrence relations that allows us to study the semilocal convergence of Newton's method under mild differentiability conditions on the first derivative of the operator. In this way, Kantorovich's theory for Newton's method has been considerably broadened.

Logroño, La Rioja, Spain   José Antonio Ezquerro Fernández
October 2019   Miguel Ángel Hernández Verón

Contents

The Newton-Kantorovich Theorem

1

Solving nonlinear equations is one of the mathematical problems that is frequently encountered in diverse scientific disciplines. Thus, with the notation

$$f(x) = 0,$$

we include the problem of finding unknown quantity x, which can be a real or complex number, a vector, a function, etc., from data provided by the function f, which can be, for example, a real function, a system of equations, a differential equation, an integral equation, etc. Even when f is a real function of a real variable, it is well known that in general it is not possible to solve a nonlinear equation accurately. Instead, iterative techniques are usually employed to obtain approximations of a solution. Among the iterative techniques, Newton's method is undoubtedly the most studied and used in practice. Thus, in order to approximate a solution α of a nonlinear equation $f(x) = 0$, Newton's method constructs, starting from an initial approximation x_0 of α, a sequence of the form

$$x_{n+1} = x_n - \frac{f(x_n)}{f'(x_n)}, \quad n \geq 0. \tag{1.1}$$

Under adequate conditions, the sequence (1.1) converges to the solution α.

Among researchers, it is customary to baptise their discoveries with their own names or with the name of a relevant celebrity in the matter. In the present, the name of the method is linked to the eminent British scientist Isaac Newton. His works at the end of the seventeenth century seem to be the germ of the method that currently bears his name. However, as it is shown in more detail in Sect. 1.1 and references therein, the method is the

© The Editor(s) (if applicable) and The Author(s), under exclusive licence to Springer Nature Switzerland AG 2020
J. A. Ezquerro Fernández, M. Á. Hernández Verón, *Mild Differentiability Conditions for Newton's Method in Banach Spaces*, Frontiers in Mathematics,
https://doi.org/10.1007/978-3-030-48702-7_1

fruit of the contributions of a great number of scientists, both before and after Newton's work. The various forms constructions that the method admits constitute another example of the plurality of backgrounds on Newton's method, see [30].

In the mid-twentieth century, the Soviet mathematician Leonid Vitaliyevich Kantorovich extended the study of Newton's method to equations defined in Banach spaces, initially what is currently known as Kantorovich's theory. Combining techniques from functional analysis and numerical analysis, Kantorovich's theory allows us to address numerous nonlinear problems such as solving integral equations, ordinary and partial differential equations, or problems of variational calculus, as it will be detailed throughout this monograph.

1.1   Brief History of Newton's Method

The "paternity" of Newton's method is attributed to Isaac Newton, who described it in several of his works published at the end of the seventeenth century. However, the idea of finding an unknown amount through successive approximations dates back many centuries before Newton. Thus, in antique Greece, techniques to approximate irrational numbers (mostly, π) by rational numbers were known. But, even earlier, 2000 years before Christ, Mesopotamians already knew techniques to approximate the square root of a number. Relevant references are abundant. For example, in [61, p. 42–43], it highlights how the famous tablet YBC 7289 (see Fig. 1.1) from the *Yale Babylonian Collection* shows a square of 30 units of side whose diagonal displays[1] the numbers 1; 24, 51, 10 and 42; 25, 35.

Conversion to the decimal system of the first number is $1.4142129629\ldots$, which matches $\sqrt{2} = 1.4142135623\ldots$ up to the fifth decimal digit. The second number is the product of 30 by the first and is, therefore, the length of the diagonal of the square. So, it seems clear that the Babylonians knew an approximate value for $\sqrt{2}$ and used it in calculations.

Another indication of the Babylonians knew how to approximate irrational numbers appears in tablet VAT6598, which is preserved in the Berlin Museum and is dated in 2000–1700 BC, where the problem of finding the diagonal of a rectangle of height 40 and side 10 is stated among others. In the current notation, the problem amounts to finding

$$\sqrt{40^2 + 10^2} = \sqrt{1700}.$$

---

[1] The Babylonians used a system of cuneiform numbering with sexadecimal base. Currently, experts in the field write the Babylonian numbers by using a mixture of our notation in base 10 and their notation in base 60. The Babylonian equivalent of the decimal comma is denoted by a point and coma. The rest of the digits are separated by commas. So, the number 5, 51, 13; 2, 30 means $5 \times 60^2 + 51 \times 60 + 13 + 2 \times 1/60 + 30 \times 1/60^2 \simeq 21073.0416$.

Fig. 1.1 Tablet YBC 7289 of
the *Yale Babylonian Collection*
(photograph of Bill Casselman)

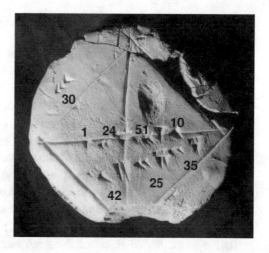

In the same tablet, the number $41; 15 = 41 + 15/60$ is proposed as an approximation. It is not known how this number was obtained or if there is evidence of the use of an iterative method, but some authors [19] mention the fact that this number coincides with the known approximation for a square root

$$\sqrt{h^2 + l^2} \simeq h + \frac{l^2}{2h}$$

for $h = 40$ and $l = 10$.

The proceeding approximation is known as the *formula of Heron* for the calculation of square roots, in which, starting from an initial approximation a of $\sqrt{A}$, the value $(a + A/a)/2$ is proposed as a new approximation. Indeed, for $A = h^2 + l^2$ and $a = h$, the approximation given in the Babylonian tablet coincides with Heron's. Although there are people who attributed the formula of Heron to the Pythagorean Archytas of Tarentum (428–347 BC) or even to Archimedes (282–212 BC), the method appears in the first volume of the *Metrica* that Heron published in the first century. This book, discovered by H. Schöne in 1896 (see [19] for details) shows how Heron estimated the area of a triangle of sides 7, 8 and 9 units, namely $\sqrt{720}$. In the same book, Heron mentions explicitly that a given approximation can be chosen as starting point to obtain best approximations. It seems clear, therefore, that this book contains the first reference of the use of an iterative method.

Now then, was Heron's method original in his time? or was it a technique already known and used by previous civilizations? The answer is in the air, although the majority of researchers of this part of the history of mathematics seem to lean towards the second option, since there is evidence of the use of Babylonian texts by mathematicians and astronomers contemporary with Heron. For example, in the work known as the *Almagest*, Claudius Ptolemy (100–170 AC) cites astronomical data of the time of the Assyrian King Nabonassar (747 BC).

From Heron's formula on techniques for calculating the square root of a number (and, in general, n-th roots) were transmitted and/or rediscovered over centuries and civilizations the seventeenth century. Although there is not much written evidence of what took place during that long period of time, we can find some references on methods for the calculation of n-th roots [19]. We can mention, for example, the Chinese mathematics book par excellence, the *Jiuzhang suanshu*, which translates as *Nine chapters of mathematical art*. There exists a third century version, with reviews of Liu Hui (220–280 AC, approximately), which contains a collection of problems that require the calculation of square and cubic roots. Later, in the fourth century, Theon of Alexandria (335–405 AC, approximately), father of Hypatia, developed a completely geometric method to approximate square roots. In the works of the Persian mathematician Sharaf Al-Din Al-Tusi (1135–1213) one finds the solutions, both algebraic and numerical, of some cubic equations. It seems that Al-Tusi was also the first to calculate the derivative of a third-degree polynomial.

In the work *Raf al-Hijab* of the Arab mathematician Al-Marrakushi Ibn Al-Banna (1256–1321), which one can translate by *Lifting the Veil*, it is shown how to calculate square roots by using series and continued fractions. It seems that Al-Banna was a great collector of the mathematical knowledge of his time, and he shows its versions of the works of earlier Arab mathematicians in his writings.

The problem of finding the n-th root of a number continued to evolve towards the more general problem of finding the roots of a polynomial equation and, even of a transcendental equation (for example, Kepler's equation). Starting with the fifteenth century, the problem bifurcated into several lines (algebraic solutions of polynomial equations, approximate solutions by using fixed-point iterations, approximations by continuous fractions, etc.). A detailed analysis of the historical development of these problems is beyond the scope of this monograph, so we refer the interested reader to one of the specialized textbooks, such as [19], or the paper [79].

Focusing on the birth of Newton's method, we can highlight the antecedent work of the French mathematician François Viète (1540–1603), who developed an ambitious project aimed at positive solutions of polynomial equations of degree 2 to 6 of generic form. Viète was the first to represent the parameters of an equation by letters, not only the unknowns. The method employed by Viète ("specious logistic" or "art of the calculation on symbols") was rooted in the Greek geometric tradition. The method of Viète, written in an archaic language and with tedious notations, did not have continuation, soon become ignored and was displaced by the Cartesian geometry. However, Viète was the first to understand the relationship between roots and coefficients of a polynomial and to try to use algebra.

It seems that the work of Viète was what inspired Isaac Newton (1643–1727) to develop his method of solving equations. The first written reference to Newton's method is found in *De analysi per aequationes numero terminorum infinitas*, in a letter written to his colleagues Barrow and Collins in 1669, which however was not published til 1711. Two years after writing this letter, in 1671, Newton developed his method in *De metodis fluxionum et serierum infinitarum*. Again, the publication of this work was delayed and it was not til 1736 that a translation was published under the title *Method of Fluxions*.

To get an idea of how Newton worked, we can illustrate his method with the same example that he considered, the equation $x^3 - 2x - 5 = 0$. Newton argued as follows:

By estimation, we see that the solution is near 2. Taking $x = 2 + \varepsilon$ and substituting in the equation, we obtain:

$$\varepsilon^3 + 6\varepsilon^2 + 10\varepsilon - 1 = 0. \tag{1.2}$$

Ignoring the terms $\varepsilon^3 + 6\varepsilon^2$ because ε is small, we have $10\varepsilon - 1 \simeq 0$, or $\varepsilon = 0.1$. Then, $x = 2.1$ is a better approximation of the solution than the initial one. Doing now $\varepsilon = 0.1 + \nu$ and substituting in (1.2), we get

$$\nu^3 + 6.3\nu^2 + 11.23\nu + 0.061 = 0.$$

Ignoring again the terms in ν of degree greater than or equal to two, we have $\nu \simeq -0.054$ and, therefore, $x = 2.046$ is an approximation that improves the previous ones. Newton indicated that the process can be repeated as many times as necessary.

Thus Newton's idea consists of adding a correcting term to a given initial approximation. To obtain this approximation, we truncate Newton's binomial at the second term in expressions of the type

$$(a + \varepsilon)^n \simeq a^n + na^{n-1}\varepsilon.$$

So, to obtain the approximate value of ε, we only have to solve a linear equation.

Writing the problem in modern notation and denoting $p(x) = x^3 - 2x - 5$, we see that the new approximation is

$$2 - \frac{p(2)}{p'(2)} = 2 + \frac{1}{10} = 2.1,$$

which corresponds to the well-known formulation of Newton's method (1.1) when $f(x)$ is the polynomial $p(x)$. However, there is no evidence that Newton used differential calculus or that he expressed the process as an iterative method in the sense that one approximation can be considered as the starting point of the next approximation. Furthermore, Newton used "his method" only to solve polynomial equations. Therefore, Newton's idea of his method is far from what we have today.

The idea of iteration is attributed to Joseph Raphson (1648–1715) (see, for example, [19, 79]), who also simplified the operational aspect of Newton's technique. In 1690, Raphson published the treatise *Analysis aequationum universalis*, in which he gave explicit formulas for the corrector term for some particular cases of equations. In particular, he calculated the corrector terms for the equations $x^3 - r = 0$ and $x^3 - px - q = 0$ and found that they are

$$\frac{r - x_0^3}{3x_0^2} \quad \text{and} \quad \frac{q + px_0 - x_0^3}{3x_0^2 - p},$$

where x_0 is the initial approximation. Notice that Raphson published his work 46 years before *Newton's Method of Fluxions*. However, Raphson was the first to recognize that Newton's method was already known in the scientific circles of that time and that his method was an improved version.

The contribution of Raphson has been historically recognized and many authors call the method the Newton-Raphson method. However, in the works of Raphson, we cannot appreciate the connection existing between the corrector term, the function that defines the equation, and its derivative.

The incorporation of the differential calculus is due to Thomas Simpson (1710–1761). As we can see in [79], Simpson, in his work *Essays on Mathematics*, published in 1740, was the one who established the method as it is currently known, except for the notational aspects (Simpson explained in a rhetoric form how to obtain the successive approximations). In addition, Simpson extended the process to arbitrary function's, not only polynomials.

On the occasion of certain observations that use infinite series, Newton seemed to be concerned with the concept of convergence, but he did not provide any solution to this problem. The first time the convergence of Newton's method is discussed in the 1768 *Traité de la résolution des équations en general* of Jean Raymond Mourraille (1720–1808). Despite the fact that it contained novel ideas, most of Mourraille's work went unnoticed.

Contrary to Newton and Raphson, Mourraille emphasized the geometric aspect of Newton's method, justifying why this method is also known as the tangent method. Mourraille used the geometric representation of Newton's method to explain the behavior of the iterative sequences it generates. Besides, Mourraille observes by first time that, depending on the starting point chosen, the sequence generated by the method can converge to any of the roots of the equation, oscillate, approach to infinity or a limit that is not a solution of the equation. Finally, Mourraille also showed that the convergence can be more or less fast, but he only indicated this in quantitative form.

Later, Joseph-Louis Lagrange (1736–1813), in his *Traité de la résolution des équations numériques de tous les degrés*, published in 1808 [40], says that the method attributed to Newton is usually employed to solve numerical equations. However, he warns that this method can be only used for equations that are already "almost solved", in the sense that a good approximation of the solution is reached. Moreover, he raises questions about the accuracy of each new iteration and observes that the method may run into difficulties in the case of multiple roots or roots that are very close to one another.

Jean Baptiste Joseph Fourier (1768–1830) was the first to analyze the rate of convergence of Newton's method in a work entitled *Question d'analyse algébraique* (1818), [40]. In this work, Fourier expressed the method in the current notation and baptized it as *la méthode newtonienne*, making explicit reference to the works of Newton, Raphson and Lagrange. Perhaps, Fourier is the "originator" of the lack of recognition for Simpson's work.

The next important mathematician to study Newton's method was Augustin Louis Cauchy (1789–1857), who started to work on it in 1821, but did not give a satisfactory formulation until the publication of the *Leçons sur le calcul différentiel* in 1829, [40]. Cauchy gave conditions, in terms of the derivatives f' and f'', to guarantee the convergence of Newton's method to a solution α of the equation $f(x) = 0$ for all starting point x_0 belonging to a certain interval. So, Cauchy was looking for results of global convergence for Newton's method, i.e., characterizing the intervals $I_\alpha \subseteq \mathbb{R}$ for which

$$\lim_{n \to \infty} x_n = \alpha, \text{ with } x_0 \in I_\alpha.$$

Although most of Cauchy's work focused on the real line, he devoted a final section to the study of complex roots. But the study of Newton's method to approximate the complex solutions of an equation contained certain surprises. Perhaps, Arthur Cayley is a mathematician who best attest to it, since in 1879 he raised the problem of characterizing the regions S_α of the complex plane for which Newton's method converges to the root α if $x_0 \in S_\alpha$. The region S_α is known as the basin of attraction of the root α. In concrete, Cayley began studying the problem of characterizing the basins of attraction for the case of a second-degree polynomial with two different roots:

$$p(z) = (z - \alpha)(z - \beta), \quad \alpha \neq \beta, \, \alpha, \, \beta \in \mathbb{C}.$$

Cayley found that the basins of attraction of the two roots were formed by the half-plans in which the complex plane is divided by the equidistant line of the two roots (the perpendicular bisector of the segment of ends α and β). If we choose a starting point in the same half-plane as a root, Newton's method converges to the root, i.e., in this case Newton's method converges to the root closest to the starting point. If we choose a starting point in the perpendicular bisector, Newton's method provides a sequence of points in the own perpendicular bisector without any apparent order, exhibiting a chaotic behavior. But the problem is complicated greatly when we consider a polynomial of third degree rather than one of second degree. Cayley's own word: "the case of a cubic equation appears to present considerable difficulties". It seems that Cayley continued working, unsuccessfully, on this problem. Eleven years later, in 1890, he again writes: "I hope to be able to apply this theory to the case of a cubic equation, but the calculations are much more difficult".

Forty years later, the works of Gaston Maurice Julia (1918) and Pierre Joseph Louis Fatou (1920) revealed that the problem to which Cayley faced was far from trivial, and essentially intractable with the knowledge and techniques of that time. Figure 1.2 shows the basins of attraction when Newton's method is applied to the cubic polynomial $p(z) = z^3 - 2z + 2$. We can then appreciate the fractal structure of the basins of attraction and understand why Cayley failed when he tried to characterize these basins. At present, it is not difficult to represent graphically the basins of attraction using various computer programs. In particular, Fig. 1.2 was produced by using *Mathematica* and following the instructions in [75].

Fig. 1.2 Bassins of attraction
of the three zeros of the
polynomial
$p(z) = z^3 - 2z + 2$

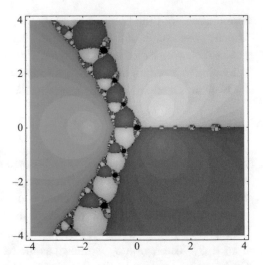

Thinking on Newton's method as a way of finding the roots of a complex function of a complex variable naturally suggests to use Newton's method to find the roots of a vector function of two variables $f : \mathbb{R}^2 \to \mathbb{R}^2$. In this case, Newton's method admits some interesting geometrical interpretations, as shown in [1, 78]. Returning to history, Simpson was the first to study a system of two transcendental equations. In the sixth of his trials, written in 1740, Simpson describes the technique to solve nonlinear systems of two equations with two unknowns by using what we call today Newton's method for systems

$$F(x) = 0, \quad \text{with} \quad F : \mathbb{R}^m \to \mathbb{R}^m.$$

In fact, Simpson indicates how to calculate the elements of the matrix $F'(x)$ for the case $m = 2$ and the solution of the corresponding system of equations $F'(x)(x_{n+1} - x_n) = -F(x_n)$ by using Cramer's rule (see [79] for more details). Simpson illustrates his technique with three examples. With the current notation, the first of them is

$$\begin{cases} y + \sqrt{y^2 - x^2} = 10, \\ x + \sqrt{y^2 + x} = 12. \end{cases}$$

Despite the innovative character of Simpson's work, it did not have much impact in his time.

The problem of the extension of Newton's method to the multidimensional case fell into oversight until the beginning of the twentieth century. In 1916, Henry Burchard Fine and Albert Arnold Bennett, both professors at Princeton University, published two articles, *On Newton's Method of Approximation* and *Newton's method in general analysis*, for systems of equations and functional equations, respectively. A convergence result for Newton's method applied to a system of equations which do not assume the existence of solution is given in the work of Fine, but it only requires conditions on the starting point. In the same work, Fine says that he did not found earlier results of this type for vector functions. In the work of Bennet, extending the work of Fine, the use of Newton's method is also justified in the case of functional equations. Both works were very innovative at a time when functional analysis was making its first steps a branch of Mathematics. In fact, only a few years later, in 1932, Stefan Banach introduced the notion of Banach spaces in his famous *Theorie des opérations linéaires* [18].

From that moment on, a number of researchers became interested in the extension of Newton's method to solve systems of equations and the publications on the topic begin to be numerous. One may consult some specific bibliographies on Newton's method ([59,60]) or search in specialized data bases.

During this period, we highlight the publications of Alexander Ostrowski [63, 64], where various convergence conditions given previously by other authors are studied and compared. In addition, error estimates are given when Newton's method is applied to solve a system of nonlinear equation.

It is in this context that Kantorovich proposed in [49] the extension of Newton's method to solve functional equations defined between two Banach spaces. Kantorovich initiated the "modern" study of Newton's method with several interesting publications, both from the theoretical and practical points of view, building what is today known as Kantorovich's theory. From the theoretical point of view, we can find numerous variants of the main result of Kantorovich's theory, called the Newton-Kantorovich theorem, that modify the hypotheses and the result (we can see some of these contributions in the compilation works [65, 77]). Kantorovich's theory found many practical applications, since a large number of problems can be solved by finding the solutions of certain nonlinear equations. In addition, Kantorovich's theory has applications in other contexts, such as optimization problems with or without constraints.

1.2   Kantorovich's Theory

As we have indicated, the generalization of Newton's method to Banach spaces is due to the Soviet mathematician and economist L. V. Kantorovich,[2] who in the middle of the twentieth century publishes several works, among which we highlight [49, 52].

The generalization of Newton's method to solve an equation

$$F(x) = 0, \tag{1.3}$$

---

[2]In addition to his contributions to the study of Newton's method, Kantorovich is well-known for his works on the theory of optimum allocation of resources, for which he was awarded with the Nobel prize in economics in 1975, jointly with Tjalling Koopmans.

where F is a nonlinear differentiable operator acting between two Banach spaces X and Y, is written in the form

$$x_{n+1} = N_F(x_n) = x_n - [F'(x_n)]^{-1} F(x_n), \quad n \geq 0, \quad \text{with } x_0 \text{ given}, \tag{1.4}$$

where $F'(x_n)$ is the first Fréchet derivative of the operator F at the point x_n and $[F'(x_n)]^{-1}$ is the inverse operator.

The most prominent contribution of Kantorovich is not the proof of his results, but the fact that one can rely on techniques of functional analysis to prove results of numerical analysis. This approach makes it possible for numerous nonlinear problems, such as the resolution of integral equations and ordinary and partial differential equations, or problems of variational calculus, can be written in the form (1.3) and to use Newton's method (1.4) to solve them to.

A remarkable feature of Kantorovich's results is that they do not assume the existence of solutions of Eq. (1.3). The type of convergence that demands conditions on the starting point x_0, but not on the solution of (1.3), is called semilocal convergence. Thus, the classical result of Kantorovich, called the Newton-Kantorovich theorem, is not only a result about the convergence of Newton's method, but also one on existence of solutions of Eq. (1.3).

Throughout the monograph, we denote $\overline{B(x, \varrho)} = \{y \in X; \|y-x\| \leq \varrho\}$ and $B(x, \varrho) = \{y \in X; \|y - x\| < \varrho\}$, and denote the set of bounded linear operators from Y to X by $\mathcal{L}(Y, X)$.

1.2.1 The Newton-Kantorovich Theorem

In this section, we present the classical Newton-Kantorovich theorem that initiated Kantorovich's theory and was published in 1948 under the following conditions on the operator F and the starting point x_0 [49]:

(K1) for $x_0 \in X$, there exists $\Gamma_0 = [F'(x_0)]^{-1} \in \mathcal{L}(Y, X)$ such that $\|\Gamma_0\| \leq \beta$;
(K2) $\|\Gamma_0 F(x_0)\| \leq \eta$;
(K3) there exists $R > 0$ such that $\|F''(x)\| \leq M$, for all $x \in B(x_0, R)$;
(K4) $h = M\beta\eta \leq \frac{1}{2}$.

Theorem 1.1 (The Newton-Kantorovich Theorem) *Let* $F : B(x_0, R) \subset X \to Y$ *be a twice continuously Fréchet differentiable operator defined on a Banach space* X *with values in a Banach space* Y. *Suppose that the conditions* (K1)-(K2)-(K3)-(K4) *are satisfied and* $B(x_0, \rho^*) \subset B(x_0, R)$ *with* $\rho^* = \frac{1-\sqrt{1-2h}}{h} \eta$. *Then, the Newton sequence defined in* (1.4) *and starting at* x_0 *converges to a solution* x^* *of the equation* $F(x) = 0$ *and the solution* x^* *and the iterates* x_n *belong to* $\overline{B(x_0, \rho^*)}$, *for all* $n \geq 0$. *Moreover, if* $h < \frac{1}{2}$, *then the solution* x^* *is unique in* $B(x_0, \rho^{**}) \cap B(x_0, R)$, *where* $\rho^{**} = \frac{1+\sqrt{1-2h}}{h} \eta$, *and, if*

$h = \frac{1}{2}$, x^* *is unique in* $\overline{B(x_0, \rho^*)}$. *Furthermore, the following error estimate holds:*

$$\|x^* - x_n\| \leq 2^{1-n} (2h)^{2^n - 1} \eta, \qquad n \geq 0. \tag{1.5}$$

Kantorovich gave two basically different proofs of the Newton-Kantorovich theorem using recurrence relations or majorant functions. The original proof was published in 1948 and uses recurrence relations [49]. Three years later, Kantorovich gave a new proof based on the concept of majorant function [52]. Both proofs are given in detail in [37].

Notice that condition (K4) of the Newton-Kantorovich theorem, often called *the Kantorovich condition*, is critical, since the value $\|F(x_0)\|$ must be small enough to ensure the starting point x_0 of Newton's method is close to a solution of the equation $F(x) = 0$.

Remark 1.2 The speed of convergence of an iterative method is usually measured by the order of convergence of the method. The first definition of an order of convergence was given in 1870 by Schröder [73], but a commonly used measure of speed of convergence in Banach spaces is the *R*-order of convergence [68], which is defined as follows:

Let $\{x_n\}$ be a sequence of points of a Banach space X that converges to a point $x^* \in X$, and let $\sigma \geq 1$ and

$$e_n(\sigma) = \begin{cases} n, & \text{if } \sigma = 1, \\ \sigma^n, & \text{if } \sigma > 1, \end{cases} \qquad n \geq 0.$$

(a) We say that σ is an *R*-order of convergence of the sequence $\{x_n\}$ if there are two constants, $b \in (0, 1)$ and $B \in (0, +\infty)$, such that

$$\|x_n - x^*\| \leq Bb^{e_n(\sigma)}.$$

(b) We say that σ is the exact *R*-order of convergence of the sequence $\{x_n\}$ if there are four constants, $a, b \in (0, 1)$ and $A, B \in (0, +\infty)$, such that

$$Aa^{e_n(\sigma)} \leq \|x_n - x^*\| \leq Bb^{e_n(\sigma)}, \qquad n \geq 0.$$

In general, check double inequalities of (b) is complicated, so that normally only seeks upper inequalities such as (a). Therefore, if we find an *R*-order of convergence σ of the sequence $\{x_n\}$, we say that the sequence $\{x_n\}$ has order of convergence at least σ. According to this terminology, estimates (1.5) guarantee that Newton's method has *R*-order of convergence at least two if $h < \frac{1}{2}$ or at least one if $h = \frac{1}{2}$ (see [68]).

1.2.2 Application to a Hammerstein Integral Equation

First of all, we note that the max norms for $\mathbb{R}^m$ and $\mathcal{C}([a, b])$ are used throughout the monograph. Moreover, in all the examples included in the monograph, we use the stopping criteria $\|x_n - x_{n-1}\| < 10^{-16}$, $n \in \mathbb{N}$, when the Newton sequence $\{x_n\}$ is calculated and

$\|x^* - x_n\| < 10^{-16}$, $n \geq 0$, for absolute errors. In addition, we also see that x^* is a good approximation of a solution of the equation under study, $F(x) = 0$, from $\|F(x^*)\| \leq$ constant $\times 10^{-16}$, so that we include the sequence $\{\|F(x_n)\|\}_{n \geq 0}$ in all the examples.

To illustrate the application of the Newton-Kantorovich Theorem 1.1, we consider a Hammerstein integral equation. Nonlinear integral equations of Hammerstein type have a strong physical background and arise, for instance, in electromagnetic fluid dynamics [69]. Here we consider nonlinear Hammerstein integral equations of the second kind [66]:

$$x(s) = u(s) + \int_a^b \mathcal{K}(s, t) \mathcal{H}(x(t)) \, dt, \quad s \in [a, b], \tag{1.6}$$

where $-\infty < a < b < +\infty$, $u(s)$ is a continuous function in $[a, b]$, the kernel $\mathcal{K}(s, t)$ is a known function in $[a, b] \times [a, b]$, $\mathcal{H}$ is a given real function and $x(s)$ is the unknown function.

If the kernel $\mathcal{K}(s, t)$ is the Green function defined by

$$\mathcal{G}(s, t) = \begin{cases} \dfrac{(b-s)(t-a)}{b-a}, & \text{if } t \leq s, \\ \dfrac{(s-a)(b-t)}{b-a}, & \text{if } s \leq t, \end{cases} \tag{1.7}$$

then equation (1.6) is equivalent to the following boundary value problem [67]:

$$\begin{cases} x''(t) + \mathcal{H}(x(t)) = 0, \\ x(a) = u(a), \ x(b) = u(b). \end{cases}$$

As Hammerstein equations of the form (1.6) cannot be solved exactly, we can use iterative methods to solve them and apply Newton's method. Moreover, we use the theoretical power of Newton's method to draw conclusions about the existence and uniqueness of a solution.

Solving equation (1.6) is equivalent to solving the equation $\mathcal{F}(x) = 0$, where $\mathcal{F} : \mathcal{C}([a, b]) \rightarrow \mathcal{C}([a, b])$ and

$$[\mathcal{F}(x)](s) = x(s) - u(s) - \int_a^b \mathcal{K}(s, t) \mathcal{H}(x(t)) \, dt, \quad s \in [a, b]. \tag{1.8}$$

Notice that $\mathcal{C}([a, b])$ is a Banach space with the max norm and so the operator (1.8) acts between two Banach spaces.

On the other hand, when we approximate a solution of an equation $F_1(\mathbf{x}) = 0$, where the operator $F_1 : \Lambda \subseteq \mathbb{R}^m \rightarrow \mathbb{R}^m$ and Λ is a nonempty open convex domain, by means of Newton's method

$$\mathbf{x}_{n+1} = \mathbf{x}_n - [F_1'(\mathbf{x}_n)]^{-1} F_1(\mathbf{x}_n), \quad n \geq 0, \quad \text{with } \mathbf{x}_0 \in \Lambda,$$

we solve at each step the system of linear equations:

$$F_1'(\mathbf{x}_n)(\mathbf{x}_{n+1} - \mathbf{x}_n) = -F_1(\mathbf{x}_n), \quad n \geq 0, \quad \text{with } \mathbf{x}_0 \in \Lambda. \tag{1.9}$$

However, if the operator F_2 involved in the equation to be solved, $F_2(x) = 0$, is of the form $F_2 : \mathcal{C}([a, b]) \rightarrow \mathcal{C}([a, b])$, we cannot use the last procedure because we do not solve the integral equation corresponding to Eq. (1.9) from operator (1.8). In addition, we cannot directly apply Newton's method, since the operator $[F_2'(x)]^{-1}$ is not known.

Then, as a first step, we use a discretization process to transform equation (1.6) into a finite-dimensional problem. To this end, we approximate the integral in (1.6) by a Gauss-Legendre quadrature formula such that

$$\int_a^b \varphi(t) \, dt \simeq \sum_{i=1}^{m} w_i \varphi(t_i),$$

where the m nodes t_i and weights w_i are known.

If we now denote the approximations of $x(t_i)$ and $u(t_i)$ by x_i and u_i, respectively, with $i = 1, 2, \ldots, m$, then Eq. (1.6) is equivalent to the following system of nonlinear equations:

$$x_i = u_i + \sum_{j=1}^{m} \alpha_{ij} \mathcal{H}(x_j), \quad j = 1, 2, \ldots, m, \tag{1.10}$$

where

$$\alpha_{ij} = w_j \, \mathcal{G}(t_i, t_j) = \begin{cases} w_j \, \dfrac{(b - t_i)(t_j - a)}{b - a}, & \text{if } j \leq i, \\[2mm] w_j \, \dfrac{(b - t_j)(t_i - a)}{b - a}, & \text{if } j > i, \end{cases} \tag{1.11}$$

if $\mathcal{K}(s, t)$ is the Green function defined in (1.7). Now, we can rewrite system (1.10) compactly in matrix form as

$$\mathbb{F}(\mathbf{x}) \equiv \mathbf{x} - \mathbf{u} - A \, \hat{\mathbf{x}} = 0, \quad \mathbb{F} : \mathbb{R}^m \longrightarrow \mathbb{R}^m, \tag{1.12}$$

where

$$\mathbf{x} = (x_1, x_2, \ldots, x_m)^T, \quad \mathbf{u} = (u_1, u_2, \ldots, u_m)^T, \quad A = (\alpha_{ij})_{i,j=1}^{m},$$

$$\hat{\mathbf{x}} = (\mathcal{H}(x_1), \mathcal{H}(x_2), \ldots, \mathcal{H}(x_m))^T.$$

Example 1.3 Using the discussion on above, let us apply Newton's method to solve the following Hammerstein equation of the form (1.6):

$$x(s) = 1 + \int_0^1 \mathcal{G}(s, t) x(t)^2 \, dt, \quad s \in [a, b], \tag{1.13}$$

where $\mathcal{G}(s, t)$ is the Green function given in (1.7) with $[a, b] = [0, 1]$.

The system (1.12) corresponding to Eq. (1.13) takes on the form

$$\mathbb{F}(\mathbf{x}) \equiv \mathbf{x} - \mathbf{1} - A \hat{\mathbf{x}} = 0, \tag{1.14}$$

where $\mathbf{u} = (1, 1, \ldots, 1)^T = \mathbf{1}, \hat{\mathbf{x}} = \left(x_1^2, x_2^2, \ldots, x_{\mathrm{m}}^2 \right)^T$ and

$$\alpha_{ij} = \begin{cases} w_j t_j (1 - t_i), & \text{if } j \leq i, \\ w_j t_i (1 - t_j), & \text{if } j > i. \end{cases}$$

In addition, the first derivative of the mapping $\mathbb{F}$ defined in (1.14) is given by

$$\mathbb{F}'(\mathbf{x}) = I - 2A \operatorname{diag}\{x_1, x_2, \ldots, x_{\mathrm{m}}\},$$

where I denotes the identity matrix. Moreover,

$$\mathbb{F}''(\mathbf{x})\mathbf{y}\mathbf{z} = (y_1, y_2, \ldots, y_{\mathrm{m}})\mathbb{F}''(\mathbf{x})(z_1, z_2, \ldots, z_{\mathrm{m}}),$$

where $\mathbf{y} = (y_1, y_2, \ldots, y_{\mathrm{m}})^T$ and $\mathbf{z} = (z_1, z_2, \ldots, z_{\mathrm{m}})^T$, so that

$$\mathbb{F}''(\mathbf{x})\mathbf{y}\,\mathbf{z} = -2A(y_1 z_1, y_2 z_2, \ldots, y_{\mathrm{m}} z_{\mathrm{m}})^T.$$

As

$$\|\mathbb{F}''(\mathbf{x})\| = \sup_{\substack{\|\mathbf{y}\|=1 \\ \|\mathbf{z}\|=1}} \|\mathbb{F}''(\mathbf{x})\mathbf{y}\,\mathbf{z}\|, \qquad \|\mathbb{F}''(\mathbf{x})\mathbf{y}\,\mathbf{z}\| \leq 2\|A\| \, \|\mathbf{y}\| \|\mathbf{z}\|,$$

it follows that the constant M that appears in Theorem 1.1 is equal to $2\|A\|$.

If we choose $\mathrm{m} = 8$ and the starting point $\mathbf{x_0} = \mathbf{1}$, which seems reasonable when $u(s) = 1$ in (1.13), Kantorovich's condition (K4) of Theorem 1.1 is satisfied because

$$h = M\beta\eta = 0.0508\ldots < \frac{1}{2}, \quad \text{where} \quad M = 0.2471\ldots, \quad \beta = 1.3134\ldots, \quad \eta = 0.1567\ldots$$

As a consequence, Newton's method converges to the solution $\mathbf{x}^* = (x_1^*, x_2^*, \ldots, x_8^*)^T$ shown in Table 1.1 after four iterations.

Table 1.1 Numerical solution $\mathbf{x}^*$ of the system (1.14)

i	x_i^*	i	x_i^*
1	1.01223942...	5	1.15980400...
2	1.05842845...	6	1.11807984...
3	1.11807984...	7	1.05842845...
4	1.15980400...	8	1.01223942...

Table 1.2 Absolute errors and $\{\|\mathbb{F}(\mathbf{x}_n)\|\}$

n	$\|\mathbf{x}^* - \mathbf{x}_n\|$	$\|\mathbb{F}(\mathbf{x}_n)\|$
0	$1.5980\ldots \times 10^{-1}$	$1.2355\ldots \times 10^{-1}$
1	$3.0824\ldots \times 10^{-3}$	$2.3565\ldots \times 10^{-3}$
2	$1.1456\ldots \times 10^{-6}$	$8.7669\ldots \times 10^{-7}$
3	$1.5742\ldots \times 10^{-13}$	$1.2048\ldots \times 10^{-13}$

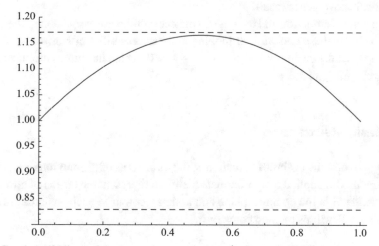

Fig. 1.3 Graph (solid line) of the approximate solution $\mathbf{x}^*$ of system (1.14)

In Table 1.2 we show the errors $\|\mathbf{x}^* - \mathbf{x}_n\|$ and the sequence $\{\|\mathbb{F}(\mathbf{x}_n)\|\}$. Examining the last sequence, we see that the vector shown in Table 1.1 is a good approximation of a solution of system (1.14) with $\mathrm{m} = 8$.

Finally, by interpolating the values of Table 1.1 and taking into account that a solution of (1.13) satisfies $x(0) = x(1) = 1$, we obtain the approximate solution drawn in Fig. 1.3. Observe also that this interpolated solution lies within the domain of existence of a solution, $\overline{B}(\mathbf{1}, 0.1609\ldots)$, obtained from the Newton-Kantorovich Theorem 1.1.

1.3 Accessibility of Newton's Method

When we study the applicability of an iterative method for solving an equation, an important aspect to consider is the set of starting points that can be chosen, so as to ensure that the iterative method converges to a solution of the equation from any point of the set. We refer to this idea as the *accessibility of an iterative method*. The accessibility of the method can be expressed in experimental terms, by means of basins of attraction and regions of accessibility, which depend on the equation to be solved, and in theoretical terms, by domains of parameters, that depend on the semilocal convergence results, and not on the equation to solve.

The basin of attraction of a solution is defined as the set of all starting points such that an iterative method converges to the solution of a particular equation with a tolerance and a maximum of iterations fixed. The region of accessibility allows us to deduce, from semilocal convergence conditions, which points can be used as starting points from which the convergence of the iterative method to a solution of a particular equation is guaranteed. Finally, the domain of parameters shows graphically, in the real plane, the relationship between the initial parameters defined from the conditions imposed on the starting points in a semilocal convergence result.

As the basins of attraction and the regions of accessibility are associated to the equations to solve, to clarify these two ways of looking at the accessibility of Newton's method, we use the simple academic complex equation $z^3 - 1 = 0$, which has three complex solutions: 1, $\exp\left(\frac{2\pi i}{3}\right)$, and $\exp\left(\frac{4\pi i}{3}\right)$.

1.3.1   Basin of Attraction

In Fig. 1.4, we show the basins of attraction of the three roots of the previous equation when Newton's method is applied to approximate them. To represent the basins of attraction, we take a rectangle $\mathfrak{R}$ in the complex plane and assign a colour to each basin of attraction of a root at which Newton's method converges.

Fig. 1.4 Bassins of attraction of the three roots of $z^3 - 1 = 0$

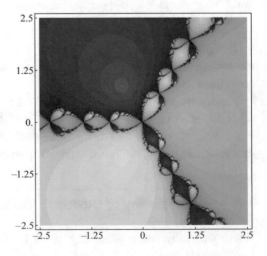

In practice, we take a grid of 512×512 points in $\mathfrak{R}$ and use these points as starting points z_0. Also, we use the rectangle $[-2.5, 2.5] \times [-2.5, 2.5]$ that contains the three roots. Newton's method starting from a point $z_0 \in \mathfrak{R}$ can converge to some of the roots or, eventually, diverge. In all the cases, we use a tolerance $\epsilon = 10^{-3}$ and a maximum of 25 iterations. We denote the three roots by $z_k^* = \exp\left(\frac{2k\pi i}{3}\right)$, $k = 0, 1, 2$. Then, we take z_0 in the rectangle and iterate Newton's method up to $\left| z_n - z_k^* \right| < \epsilon$ for $k = 0, 1, 2$. If we do not obtain the desired tolerance with 25 iterations, we stop and decide that the iterative method starting at z_0 does not converge to any root. In particular, we assign cyan, magenta and yellow for the basins of attraction of the three roots 1, $\exp\left(\frac{2\pi i}{3}\right)$, and $\exp\left(\frac{4\pi i}{3}\right)$, respectively. The colour is lighter or darker according to the number of iterations needed to reach the root with the fixed precision required. We do not assign any colour to the points z_0 of the rectangle for which Newton's method starting from z_0 does not reach any root with tolerance 10^{-3} in a maximum of 25 iterations. For other strategies, one can consult [75] and the references cited there. The graphics have been generated with Mathematica 5.1 [76].

1.3.2 Region of Accessibility

As we have indicated before, the region of accessibility shows the set of starting points that satisfy the semilocal convergence conditions required by an iterative method that is applied to approximate a solution of a particular equation.

We know that to the starting points of Newton's method there are associated the parameters β, η and M that appear in the initial conditions (K1)-(K2)-(K3) of the Newton-Kantorovich Theorem 1.1. To represent the region of accessibility of Newton's method, we colour the points whose associated parameters satisfy condition (K4) and use no colour if (K4) is not satisfied. The region of accessibility associated with a solution of an equation is then the set of starting points for which the convergence of Newton's method is guaranteed; in particular, it includes the set of starting points which satisfy the convergence conditions (K1)-(K2)-(K3)-(K4) of Theorem 1.1.

For the last, we consider, for example, the disc $B(0, 1.6)$ as the domain of the complex function $d(z) = z^3 - 1$. Then, $M = 9.6$. To paint the regions of accessibility, we take $z_0 \in B(0, 1.6)$ and colour red all the points z_0 that satisfy condition (K4), $M\beta\eta \leq \frac{1}{2}$, of Theorem 1.1. In Fig. 1.5, we see the regions of accessibility associated with the three solutions of equation $z^3 - 1 = 0$ when they are approximated by Newton's method.

1.3.3 Domain of Parameters

Apart from the empirical observation of the accessibility of Newton's method given by the basins of attraction and the regions of accessibility, we can carry out a theoretical study of

Fig. 1.5 Regions of
accessibility of the three roots
of $z^3 - 1 = 0$ associated with
the Newton-Kantorovich
Theorem 1.1

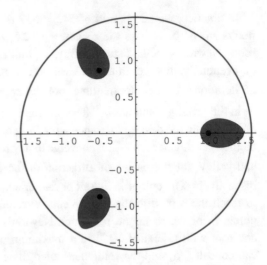

the accessibility of the method from the convergence conditions imposed in the Newton-Kantorovich Theorem 1.1.

Recall that the convergence conditions imposed in Theorem 1.1 have two distinguished parts: conditions imposed on the starting point x_0 (see (K1)–(K2)) and a condition imposed on the operator F (see (K3)). To study the restrictions that are imposed by the initial conditions, we use the domain parameters, which show graphically in the real plane the relationship between the parameters defined from the initial conditions.

Thus, if we study the accessibility of Newton's method theoretically based on Theorem 1.1, it suffices to take into account that, given a starting point $x_0 \in \Omega$, the method associates the parameters β and η that appear in (K1)-(K2) and, from condition $h = M\beta\eta \leq \frac{1}{2}$ of (K4), we can define the domain of parameters associated with the theorem as the subset of the real plane given by

$$\left\{ (x, y) \in \mathbb{R}_+^2 : xy \leq \frac{1}{2} \right\},$$

where we can choose different values for the axes of the xy-plane depending on the study that we want to do. For example, if we are interested in measuring the variation of the parameter η (i.e., the proximity measure from x_0 to x^*), we can choose the value of η on the x-axis and the value of $M\beta$ on the y-axis.

Moreover, we observe that the parameter η "measures" how well x_0 approximates a solution x^* of the equation and $\eta = 0$ if $x_0 = x^*$. On the other hand, the parameter M required for the operator F, which is defined from condition (K3), is always a fixed value, so that M does not influence the domain of parameters, since M does not change for different starting points.

Fig. 1.6 Domain of
parameters of Newton's
method associated with the
Newton-Kantorovich
Theorem 1.1

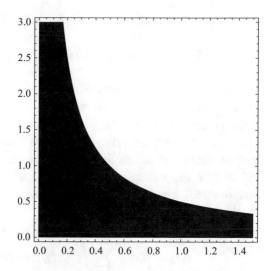

In Fig. 1.6, we show the domain of parameters associated with the Newton-Kantorovich Theorem 1.1. To represent it graphically, we consider the xy-plane and color black the values of the parameters that satisfy the condition $xy \leq \frac{1}{2}$, where the x axis represents the values of η and the y-axis the values of $M\beta$.

1.4   Mild Alternatives to Kantorovich's Conditions

If we consider $F : \Omega \subseteq X \rightarrow Y$, where Ω is a nonempty open convex domain of a Banach space X and Y is a Banach space, and apply the algorithm of Newton's method given in (1.4), we see that only the first Fréchet derivative F' of the operator F is involved. According to this, we can give only conditions on F', so that the semilocal convergence of the method is guaranteed. So, condition (K3) on $F''(x)$ can be easily replaced in the Newton-Kantorovich theorem by a Lipschitz continuity condition on F':

$$\|F'(x) - F'(y)\| \leq L\|x - y\|, \quad x, y \in \Omega. \tag{1.15}$$

This was done by several authors, the first of which was, perhaps, Fenyö [39], although the best known version is due to Ortega [62], who presented a modification of the approach given by Kantorovich in terms of majorant functions, easier to read and understand, that is based on the concept of "majorizing sequence", providing upper bounds for the errors of Newton's method.

Adopting the previous idea, we focus through the monograph on guaranteeing the semilocal convergence of Netwon's method under mild differentiability conditions on the first Fréchet derivative F' of the operator involved F. In addition to replacing condition (K3) by condition (1.15), we can also replace condition (K3) by the following generalization (see [48, 77] and references therein):

$$\|F'(x) - F'(y)\| \le K\|x - y\|^p, \quad p \in [0, 1], \quad x, y \in \Omega. \tag{1.16}$$

We then says that the operator F' is (K, p)-Hölder continuous in Ω. Observe that (1.16) reduces to (1.15) if $K = L$ and $p = 1$.

In practice, checking these conditions is difficult in some problems, since certain technical difficulties arise. In particular, we cannot analyze the convergence of the method to a solution of equations that involve sums of operators which satisfy (1.15) or (1.16), indistinctly. For example, consider the following nonlinear integral equation of mixed Hammerstein type that can be found, for instance, in dynamical models of chemical reactors [41]:

$$x(s) = u(s) + \sum_{i=1}^{N} \int_a^b \mathcal{K}_i(s, t)\mathcal{H}_i(x(t)) \, dt, \quad s \in [a, b], \tag{1.17}$$

where $-\infty < a < b < +\infty$, $u(s)$ is a continuous function in $[a, b]$, the kernels $\mathcal{K}_i(s, t)$ are known functions in $[a, b] \times [a, b]$, $\mathcal{H}_i$ are a given real functions, for $i = 1, 2, \ldots, N$, and $x(s)$ is the unknown function.

Solving Eq. (1.17) is equivalent to solving the equation $\mathcal{F}(x) = 0$ with $\mathcal{F} : \Omega \subseteq \mathcal{C}([a, b]) \to \mathcal{C}([a, b])$ and

$$[\mathcal{F}(x)](s) = x(s) - u(s) - \sum_{i=1}^{N} \int_a^b \mathcal{K}_i(s, t)\mathcal{H}_i(x(t)) \, dt, \quad s \in [a, b].$$

If we now consider an Eq. (1.17) such that $\mathcal{H}_i'(x(t))$, for $i = 1, 2, \ldots, N$, is (K_i, p_i)-Hölder continuous in Ω, then the above operator $\mathcal{F}(x)$ satisfies neither the condition (1.15) nor the condition (1.16) in Ω when, for example, the max-norm is chosen. In this case,

$$\|\mathcal{F}'(x) - \mathcal{F}'(y)\| \le \sum_{i=1}^{N} K_i\|x - y\|^{p_i}, \quad p_i \in [0, 1], \quad x, y \in \Omega,$$

so that, to solve this type of problems and relax conditions (1.15) and (1.16), we can consider the following generalization:

$$\|F'(x) - F'(y)\| \le \omega(\|x - y\|), \quad x, y \in \Omega, \tag{1.18}$$

where $\omega : [0, +\infty) \to \mathbb{R}$ is a continuous nondecreasing function such that $\omega(0) \ge 0$, [28, 29]. Thus, for the last particular case, $\omega(z) = \sum_{i=1}^{N} K_i z^{p_i}$. Moreover, conditions (1.15) and (1.16) are particular cases of (1.18), since (1.18) reduces to (1.15) and (1.16) if $\omega(z) = Lz$ and $\omega(z) = Kz^p$, respectively.

If we now focus on the type of conditions that are required to ensure the semilocal convergence of Newton's method, we highlight that we analyse the semilocal convergence of the method from conditions on the starting point x_0 and conditions on the operator involved F, along with a condition that connects the previous two types of conditions. In particular, we distinguish four types of conditions imposed on F': general conditions on the operator in the domain of definition, conditions on the operator specified at the starting point, conditions that combine the two previous types, and finally conditions centered at an auxiliary point.

1.5   A Technique Based on Recurrence Relations

In 1951, Kantorovich introduced in [52] the well-known "majorant principle" to prove the semilocal convergence of Newton's method under condition (K3) for the operator involved F. This principle was based on the concept of "majorizing sequence", that is later introduced by Ortega in [62]. We say that a sequence of scalar numbers $\{t_n\}$ *majorizes* a sequence $\{x_n\}$ of elements of a Banach space X if

$$\|x_{n+1} - x_n\| \le t_{n+1} - t_n, \qquad n \ge 0.$$

The interest in majorizing sequences is comes from the gact that the convergence of the sequence $\{x_n\}$ in the Banach space X will follow from the convergence of the scalar sequence $\{t_n\}$. Indeed, if $\{t_n\}$ converges to t^*, then there exists $x^* \in X$, such that the sequence $\{x_n\}$ converges to x^* and

$$\|x^* - x_n\| \le t^* - t_n, \qquad n \ge 0.$$

To show this, Kantorovich proved that a majorizing sequence $\{t_n\}$ is obtained by applying Newton's method to the scalar function

$$\wp(t) = \frac{M}{2}t^2 - \frac{1}{\beta}t + \frac{\eta}{\beta}, \qquad (1.19)$$

i.e.,

$$t_{n+1} = N_{\wp}(t_n) = t_n - \frac{\wp(t_n)}{\wp'(t_n)}, \quad n \ge 0, \quad \text{with } t_0 = 0.$$

The technique based on majorizing sequences is the most used technique to prove the semilocal convergence of Newton's method.

Note that the function $\wp$ defined in (1.19) can be easily obtained by an interpolation procedure fitting from fixing the three coefficients of the quadratic polynomial by conditions (K1)-(K2)-(K3) of the Newton-Kantorovich Theorem 1.1. However, if we replace condition (K3) by a milder one (afor example, considering operators whose first derivative satisfies condition (1.16) or condition (1.18), etc.), the construction of a scalar function, such as $\wp$, is not obvious, and in fact can be very complicated, so that obtaining a majorizing sequence of the previous type is difficult. To deal with this situation, we have recently developed an alternative technique that simplifies the analysis of the semilocal convergence of Newton's method under mild differentiability conditions on the first derivative of the operator involved. This technique relies on the construction of a scalar sequence, not majorizing, that satisfies a system of recurrence relations, which enables us to guarantee the convergence of Newton's method in a Banach space. The application is simple and has certain advantages over the usual technique of majorizing sequences. On the one hand, we can generalize the results obtained under conditions of Newton-Kantorovich type and, on the other hand, the results obtained through majorizing sequences can be improved. In addition, we can obtain a priori error estimates ([29]). The behavior of this technique is optimal when the conditions on the operator F are relaxed and, in particular, when condition (K3) of the Newton-Kantorovich theorem is relaxed. Moreover, this technique can be extended to any iterative method [26, 27, 47].

Operators with Lipschitz Continuous First Derivative

2

Examining the algorithm of Newton's method,

$$x_{n+1} = x_n - [F'(x_n)]^{-1} F(x_n), \quad n \geq 0, \quad \text{with } x_0 \text{ given,}$$

we see that it involves only the operator F and its first Fréchet derivative F', suggests that trying to impose conditions only on the operators F and F' to guarantee the convergence of Newton's method. Thus, the continuity condition (K3) on $F''(x)$ in the Newton-Kantorovich Theorem 1.1 can be easily replaced by a Lipschitz continuity condition on F', i.e.,

$$\|F'(x) - F'(y)\| \leq L\|x - y\|, \quad x, y \in \Omega. \tag{2.1}$$

Moreover, condition (K3) in Theorem 1.1 implies condition (2.1) in a given open ball.

Adopting this approach, we will establish the semilocal convergence of Newton's method under condition (2.1). For this, we introduce in this chapter a technique to prove the convergence of the method that is based on recurrence relations. Then, we define a scalar sequence from the convergence conditions required and prove an associated system of recurrence relations. Finally, we apply this type of analysis to a Chandrasekhar integral equation that arises in the study of isotropic scattering, draw conclusions about the existence and uniqueness of solution, and approximate numerically a solution of the equation by Newton's method.

© The Editor(s) (if applicable) and The Author(s), under exclusive licence to Springer Nature Switzerland AG 2020
J. A. Ezquerro Fernández, M. Á. Hernández Verón, *Mild Differentiability Conditions for Newton's Method in Banach Spaces*, Frontiers in Mathematics,
https://doi.org/10.1007/978-3-030-48702-7_2

2.1   Convergence Analysis

Notice that Kantorovich's conditions (K1)-(K2)-(K3)-(K4) of Theorem 1.1 can be classified into three types: two conditions on the starting point x_0 (conditions (K1)-(K2)), a condition on the operator F (condition (K3)) and a condition that combines the last three conditions (condition (K4)). Taking into account this classification, we rewrite conditions (K1)-(K2)-(K3)-(K4) in the following way:

- (A1) For a point x_0 there exist $\Gamma_0 = [F'(x_0)]^{-1} \in \mathcal{L}(Y, X)$, with $\|\Gamma_0\| \leq \beta$ and $\|\Gamma_0 F(x_0)\| \leq \eta$.
- (A2) There exists a constant $L \geq 0$ such that $\|F'(x) - F'(y)\| \leq L\|x - y\|$ for $x, y \in \Omega$.
- (A3) $a_0 = L\beta\eta \leq \frac{1}{2}$ and $B(x_0, R) \subset \Omega$, where $R = \frac{2(1-a_0)}{2-3a_0}\eta$.

Now the Newton-Kantorovich Theorem 1.1 can be restated as follows.

Theorem 2.1 *Let $F : \Omega \subseteq X \to Y$ be a continuously Fréchet differentiable operator defined on a nonempty open convex domain Ω of a Banach space X with values in a Banach space Y. Suppose that conditions (A1)-(A2)-(A3) are satisfied. Then, the Newton sequence $\{x_n\}$ starting at x_0 converges to a solution x^* of the equation $F(x) = 0$. Moreover, $x_n, x^* \in \overline{B(x_0, R)}$ and x^* is unique in $B(x_0, \eta/a_0) \cap \Omega$. The Newton sequence $\{x_n\}$ has R-order of convergence at least two if $a_0 \in (0, 1/2)$, and at least one if $a_0 = 1/2$, and*

$$\|x^* - x_n\| \leq \left(\gamma^{2^n - 1}\right) \frac{\Delta^n}{1 - \gamma^{2^n}\Delta}\eta, \quad n \geq 0, \tag{2.2}$$

where $\gamma = \frac{a_0}{2(1-a_0)^2}$ and $\Delta = 1 - a_0$.

To prove Theorem 2.1, we introduce below a technique based on recurrence relations. Namely, we construct, using the parameters β, η and L, a system of two recurrence relations involving a sequence of positive real numbers, whose analysis guarantees the convergence of the Newton sequence $\{x_n\}$ in the Banach space X.

2.1.1   Recurrence Relations

Start with $a_0 = L\beta\eta$ and define the scalar sequence $\{a_n\}$ by the rule

$$a_{n+1} = a_n f(a_n)^2 g(a_n), \quad n \geq 0, \tag{2.3}$$

where

$$f(t) = \frac{1}{1-t} \quad \text{and} \quad g(t) = \frac{t}{2}. \tag{2.4}$$

Since an obvious problem results if $a_0 = 0$, we take $a_0 > 0$.

We first prove that for the sequences (2.3) and $\{x_n\}$, the following recurrence relations (inequalities) hold:

$$\|\Gamma_1\| = \|[F'(x_1)]^{-1}\| \le f(a_0)\|\Gamma_0\|, \tag{2.5}$$

$$\|x_2 - x_1\| \le f(a_0)g(a_0)\|x_1 - x_0\|. \tag{2.6}$$

To this end, we assume that

$$x_1 \in \Omega \quad \text{and} \quad a_0 < 1. \tag{2.7}$$

Then,

$$\|I - \Gamma_0 F'(x_1)\| \le \|\Gamma_0\|\|F'(x_0) - F'(x_1)\| \le \beta L \|x_1 - x_0\| \le L\beta\eta = a_0 < 1.$$

Therefore, by the Banach lemma on invertible operators, the operator Γ_1 exists and

$$\|\Gamma_1\| \le \frac{\|\Gamma_0\|}{1 - \|I - \Gamma_0 F'(x_1)\|} \le f(a_0)\|\Gamma_0\|.$$

Since $\{x_n\}$ is a Newton sequence, the Taylor series expansion yields

$$F(x_1) = F(x_0) + F'(x_0)(x_1 - x_0) + \int_{x_0}^{x_1} \left(F'(x) - F'(x_0)\right) dx$$

$$= \int_{x_0}^{x_1} (F'(x) - F'(x_0)) \, dx$$

$$= \int_0^1 (F'(x_0 + t(x_1 - x_0)) - F'(x_0))(x_1 - x_0) \, dt,$$

$$\|F(x_1)\| \le \frac{L}{2}\|x_1 - x_0\|^2 \le \frac{L\eta}{2}\|x_1 - x_0\|.$$

Consequently,

$$\|x_2 - x_1\| = \|\Gamma_1 F(x_1)\| \le \|\Gamma_1\|\|F(x_1)\| \le f(a_0)g(a_0)\|x_1 - x_0\|.$$

Moreover, if $B(x_0, R) \subset \Omega$ and $f(a_0)g(a_0) < 1$, then $\|x_2 - x_1\| < \|x_1 - x_0\|$, and so

$$\|x_2 - x_0\| \le \|x_2 - x_1\| + \|x_1 - x_0\|$$

$$\le (1 + f(a_0)g(a_0))\|x_1 - x_0\|$$

$$< \frac{\eta}{1 - f(a_0)g(a_0)}$$

$$= \frac{2(1 - a_0)}{2 - 3a_0} \eta$$

$$= R \tag{2.8}$$

and $x_2 \in \Omega$.

2.1.2   Analysis of the Scalar Sequence

Our next aim is to analyze the scalar sequence (2.3) so that we can guarantee the convergence of the Newton sequence $\{x_n\}$ in the Banach space X. For this, we just need to show $\{x_n\}$ is a Cauchy sequence. First, we give a technical lemma whose simple proof is omitted.

Lemma 2.2 *Let f and g be the two real functions defined in (2.4). Then*

(a) *f is increasing and $f(t) > 1$ in $(0, 1)$;*
(b) *g is increasing;*
(c) *for $\gamma \in (0, 1)$, we have $f(\gamma t) < f(t)$ if $t > 0$ and $g(\gamma t) = \gamma g(t)$.*

Now, we prove some properties of the scalar sequence defined in (2.3).

Lemma 2.3 *Let f and g be the two scalar functions defined in (2.4). If $a_0 \in (0, 1/2)$, then*

(a) *$f(a_0)^2 g(a_0) < 1$;*
(b) *the sequence $\{a_n\}$ is strictly decreasing;*
(c) *$a_n < 1$, for all $n \geq 0$.*

If $a_0 = 1/2$, then $a_n = a_0 < 1$ for all $n \geq 1$.

Proof First, consider the case $a_0 \in (0, 1/2)$. Item (a) follows immediately. Item (b) is proved by induction on n. As $f(a_0)^2 g(a_0) < 1$, it is clear that $a_1 < a_0$. If we now suppose that $a_j < a_{j-1}$, for $j = 1, 2, \ldots, n$, then

$$a_{n+1} = a_n f(a_n)^2 g(a_n) < a_n f(a_0)^2 g(a_0) < a_n,$$

since f and g are increasing; hence, the sequence $\{a_n\}$ is strictly decreasing. In addition, $a_n < 1$, for all $n \geq 1$, so item (c) is proved as well.

Second, if $a_0 = 1/2$, then $f(a_0)^2 g(a_0) = 1$ and it is obvious that $a_n = a_0 = 1/2 < 1$, for all $n \geq 0$. ∎

Lemma 2.4 *Let f and g be the two real functions given in (2.4). If $a_0 \in (0, 1/2)$ and $\gamma = \frac{a_1}{a_0}$, then*

(a) $a_n < \gamma^{2^{n-1}} a_{n-1}$ *and* $a_n < \gamma^{2^n-1} a_0$, *for all $n \geq 2$,*
(b) $f(a_n)g(a_n) < \gamma^{2^n-1} f(a_0)g(a_0) = \gamma^{2^n}/f(a_0)$, *for all $n \geq 1$.*

If $a_0 = 1/2$, then $f(a_n)g(a_n) = f(a_0)g(a_0) = 1/f(a_0)$, for all $n \geq 1$.

Proof We first consider the case $a_0 \in (0, 1/2)$. We prove item (a) by induction on n. If $n = 2$, Lemma 2.2(b) shows that

$$a_2 = a_1 f(a_1)^2 g(a_1) = \gamma a_0 f(\gamma a_0)^2 g(\gamma a_0) < \gamma^2 a_1 = \gamma^3 a_0.$$

Now suppose

$$a_{n-1} < \gamma^{2^{n-2}} a_{n-2} < \gamma^{2^{n-1}-1} a_0,$$

then, by the same reasoning,

$$
\begin{aligned}
a_n &= a_{n-1} f(a_{n-1})^2 g(a_{n-1}) \\
&< \gamma^{2^{n-2}} a_{n-2} f\left(\gamma^{2^{n-2}} a_{n-2}\right)^2 g\left(\gamma^{2^{n-2}} a_{n-2}\right) \\
&< \gamma^{2^{n-1}} a_{n-1} \\
&< \gamma^{2^{n-1}} \gamma^{2^{n-2}} a_{n-2} \\
&< \cdots < \gamma^{2^n-1} a_0.
\end{aligned}
$$

To prove item (b), we observe that for $n \geq 1$,

$$f(a_n)g(a_n) < f\left(\gamma^{2^n-1} a_0\right) g\left(\gamma^{2^n-1} a_0\right) < \gamma^{2^n-1} f(a_0)g(a_0) = \frac{\gamma^{2^n}}{f(a_0)}, \quad n \geq 1.$$

The case $a_0 = 1/2$ is dealt with analogously. ∎

2.1.3 Proof of the Semilocal Convergence

We are now ready to prove the semilocal convergence result given in Theorem 2.1 for Newton's method when the method is applied to operators with Lipschitz continuous first Fréchet derivative in Ω.

Proof We start with the case $a_0 \in (0, 1/2)$. First, we prove that sequence $\{x_n\}$ enjoys the following properties for $n \geq 2$:

(I$_n$)   There exists the operator $\Gamma_{n-1} = [F'(x_{n-1})]^{-1}$ and $\|\Gamma_{n-1}\| \leq f(a_{n-2})\|\Gamma_{n-2}\|$.

(II$_n$) $\|x_n - x_{n-1}\| \leq f(a_{n-2})g(a_{n-2})\|x_{n-1} - x_{n-2}\|$.

(III$_n$) $x_n \in \Omega$.

Notice that $x_1 \in \Omega$, since $\eta < R$. Then, (2.5), (2.6) and (2.8) imply the items (I$_2$)-(II$_2$)-(III$_2$). If we now suppose the items (I$_{n-1}$)-(II$_{n-1}$) and proceed as we did when we proved the items (I$_2$)-(II$_2$), one easily obtains the (I$_n$)-(II$_n$). Thus, by induction, we prove that the items (I$_{n-1}$)-(II$_{n-1}$) are true for all positive integers n. Note that $a_n < 1$ for all $n \geq 0$ and let us prove item (I$_n$). By item (II$_n$) and Lemma 2.4 (b),

$$\|x_n - x_0\| \leq \sum_{i=0}^{n-1} \|x_{i+1} - x_i\|$$

$$\leq \left(1 + \sum_{i=0}^{n-2}\left(\prod_{j=0}^{i} f(a_j)g(a_j)\right)\right) \|x_1 - x_0\|$$

$$< \left(1 + \sum_{i=0}^{n-2}\left(\prod_{j=0}^{i} f(a_0)g(a_0)\gamma^{2^j - 1}\right)\right) \|x_1 - x_0\|$$

$$= \left(1 + \sum_{i=0}^{n-2}\left(\prod_{j=0}^{i} \left(\gamma^{2^j}\Delta\right)\right)\right) \|x_1 - x_0\|$$

$$= \left(1 + \sum_{i=0}^{n-2}\left(\gamma^{2^{1+i}-1}\Delta^{1+i}\right)\right) \|x_1 - x_0\|,$$

where $\gamma = \frac{a_1}{a_0} < 1$ and $\Delta = f(a_0)g(a_0)/\gamma = 1/f(a_0) = 1 - a_0 < 1$. Next, Bernoulli's inequality $((1 + z)^n - 1 \geq nz$ if $z > -1)$ implies that $\gamma^{2^{1+i}-1} = \gamma^{2(2^i-1)+1} \leq \gamma^{2i+1}$ and, as a consequence,

$$\|x_n - x_0\| < \left(1 + \gamma\Delta\sum_{i=0}^{n-2}\gamma^{2i}\Delta^i\right) \|x_1 - x_0\|$$

$$< \left(1 + \gamma\Delta\frac{1 - (\gamma^2\Delta)^{n-1}}{1 - \gamma^2\Delta}\right) \eta$$

$$< \frac{\eta}{1 - \gamma\Delta}$$

$$= \frac{2(1 - a_0)}{2 - 3a_0}\eta$$

$$= R.$$

Therefore, $x_n \in B(x_0, R)$ and, as $B(x_0, R) \subset \Omega$, we have $x_n \in \Omega$, for all $n \geq 0$. Note that the conditions (2.7) are now satisfied for all x_n and a_{n-1} with $n \geq 2$.

Second, we prove that $\{x_n\}$ is a Cauchy sequence. For this, we proceed in much the same way as above. So, for $m \geq 1$ and $n \geq 1$, we have

$$\|x_{n+m} - x_n\| \leq \sum_{i=n}^{n+m-1} \|x_{i+1} - x_i\|$$

$$\leq \left(1 + \sum_{i=n}^{n+m-2} \left(\prod_{j=n}^{i} f(a_j)g(a_j)\right)\right) \|x_{n+1} - x_n\|$$

$$\leq \sum_{i=n-1}^{n+m-2} \left(\prod_{j=0}^{i} f(a_j)g(a_j)\right) \|x_1 - x_0\|$$

$$< \sum_{i=n-1}^{n+m-2} \left(\prod_{j=0}^{i} \gamma^{2^j-1} f(a_0)g(a_0)\right) \|x_1 - x_0\|$$

$$= \sum_{i=n-1}^{n+m-2} \left(\prod_{j=0}^{i} \left(\gamma^{2^j} \Delta\right)\right) \|x_1 - x_0\|$$

$$= \sum_{i=n-1}^{n+m-2} \left(\gamma^{2^{1+i}-1} \Delta^{1+i}\right) \|x_1 - x_0\|$$

$$= \sum_{i=0}^{m-1} \left(\gamma^{2^{n+i}-1} \Delta^{n+i}\right) \|x_1 - x_0\|$$

from item (II_n) and Lemma 2.4(b), and, applying the Bernoulli inequality, $\gamma^{2^{n+i}-1} = \gamma^{2^n-1}\gamma^{2^n(2^i-1)} \leq \gamma^{2^n-1}\gamma^{2^n i}$, we see that

$$\|x_{n+m} - x_n\| < \left(\sum_{i=0}^{m-1} \left(\gamma^{2^n i} \Delta^i\right)\right) \gamma^{2^n-1} \Delta^n \eta < \frac{1 - \left(\gamma^{2^n} \Delta\right)^m}{1 - \gamma^{2^n} \Delta} \gamma^{2^n-1} \Delta^n \eta. \tag{2.9}$$

Thus, $\{x_n\}$ is a Cauchy sequence, and so there exists $x^* \in \overline{B(x_0, R)}$ such that $x^* = \lim_n x_n$.

Third, to prove that x^* is a solution of the equation $F(x) = 0$, we argue in the usual way. Since $\|\Gamma_n F(x_n)\| \to 0$, as $n \to \infty$, and since $\|F(x_n)\| \leq \|F'(x_n)\|\|\Gamma_n F(x_n)\|$ and the sequence $\{\|F'(x_n)\|\}$ is bounded,

$$\|F'(x_n)\| \leq \|F'(x_0)\| + L\|x_n - x_0\| < \|F'(x_0)\| + LR,$$

we conclude that $\|F(x_n)\| \to 0$ as $n \to \infty$. As a consequence, we obtain $F(x^*) = 0$ by the continuity of F in $B(x_0, R)$.

Fourth, to prove the uniqueness of the solution x^*, suppose that y^* is another solution of $F(x) = 0$ in $B(x_0, \eta/a_0) \cap \Omega$. Then, the approximation formula

$$0 = \Gamma_0(F(y^*) - F(x^*)) = \left(\int_0^1 \Gamma_0 F'(x^* + \tau(y^* - x^*)) \, d\tau \right)(y^* - x^*) = J(y^* - x^*),$$

implies that $x^* = y^*$, provided that the operator $\displaystyle\int_0^1 \Gamma_0 F'(x^* + \tau(y^* - x^*)) \, d\tau = J$ is invertible. But

$$\|I - J\| \le \|\Gamma_0\| \int_0^1 \|F'(x^* + \tau(y^* - x^*)) - F'(x_0)\| \, d\tau$$

$$\le L\beta \int_0^1 \|x^* + \tau(y^* - x^*) - x_0\| \, d\tau$$

$$\le L\beta \int_0^1 \left((1 - \tau)\|x^* - x_0\| + \tau\|y^* - x_0\|\right) d\tau$$

$$\le L\beta \int_0^1 (R(1 - \tau) + (\eta/a_0)\tau) \, d\tau$$

$$< 1,$$

so the operator J^{-1} exists by the Banach lemma on invertible operators.

Finally, letting $m \to \infty$ in (2.9), we obtain (2.2), for all $n \ge 0$. Moreover, (2.2) implies that the R-order of convergence of the Newton sequence $\{x_n\}$ is at least two (see [48]).

On the other hand, if $a_0 = 1/2$, then $a_n = a_0 = 1/2$ for all $n \ge 0$. Then, proceeding by analogy with the case where $a_0 \in (0, 1/2)$, we obtain the same results by taking now into account that $\gamma = 1$ and $\Delta = f(a_0)g(a_0) < 1$, except for the R-order of convergence, which in this case is at least one. ∎

Remark 2.5 Notice that inequalities (I_n)-(II_n) given in the proof of Theorem 2.1 reduce to equalities when the equation to be solved is

$$\zeta(t) = \frac{L}{2}t^2 - \frac{t}{\beta} + \frac{\eta}{\beta} = 0,$$

so that (I_n)-(II_n) are optimal in this case. Note that the polynomial ζ is just the quadratic polynomial of Kantorovich usually employed to construct a majorizing sequence and prove the semilocal convergence of Newton's method if condition (K3) is replaced by condition (A2), see [62].

Since estimates regarding consecutive points are the optimal tool to measure $\|x^* - x_n\|$, we can look for an element x_j $(j > n)$ of the sequence $\{x_n\}$ such that $\|x^* - x_j\|$ is small enough and then estimate $\|x^* - x_n\|$ by

$$\|x^* - x_n\| \le \|x^* - x_j\| + \|x_j - x_{j-1}\| + \cdots + \|x_{n+1} - x_n\|, \quad j > n, \quad n \ge 1,$$

so that the errors given in Eq. (2.2) of Theorem 2.1 are improved.

2.2   Application to a Chandrasekhar Integral Equation

The solutions for the problems of constant net flux and diffuse reflection obtained under conditions of isotropic scattering have revealed a relationship between the two problems (see [20]). In isotropic scattering it is possible to go back to the original transfer equations and derive the indicated relationship as an integral of the equations. But when more general scattering laws are considered, the relationship which emerge are so involved that the attempt to establish them as integrals of the relevant equations would hardly reveal their physical meaning. It is therefore of interest to observe that the real origin of the relationship between the two problems has to be traced. The invariance principles play an important role (see [20]) because they can be used to derive integral equations for the functions governing the angular distribution of radiation in various problems. These integral equations are in general nonlinear.

Example 2.6 An explicit form of the integral equations in the case of isotropic scattering is the Chandrasekhar equation

$$x(s) = 1 + \lambda s\, x(s) \int_0^1 \frac{x(t)}{s+t}\, dt, \quad s \in [0, 1], \tag{2.10}$$

where λ is the albedo for single scattering. Equation (2.10) arises also in the study of radiative transfer, neutron transport and the kinetic theory of the gases.

Next, we use Theorem 2.1 to draw conclusions about the existence and uniqueness of a solution of Eq. (2.10) with $\lambda = \frac{1}{4}$, which is used by other authors [5, 6, 20], and approximate a solution by Newton's method.

In terms of operators, solving Eq. (2.10) with $\lambda = 1/4$ is equivalent to solving the equation $\mathcal{F}(x) = 0$ with $\mathcal{F} : C([a, b]) \to C([a, b])$ given by

$$[\mathcal{F}(x)](s) = x(s) - 1 - \frac{1}{4} s\, x(s) \int_0^1 \frac{x(t)}{s+t}\, dt, \quad s \in [0, 1].$$

Now, to apply Theorem 2.1, we first observe that for $s = 0$ Eq. (2.10) becomes $x(0) = 1$, so that a reasonable choice of starting point for Newton's method is $x_0(s) = 1$ for all

$s \in [0, 1]$. In addition,

$$[\mathcal{F}'(x)y](s) = y(s) - \frac{s}{4}x(s) \int_0^1 \frac{y(t)}{s+t} \, dt - \frac{s}{4}y(s) \int_0^1 \frac{x(t)}{s+t} \, dt, \quad s \in [0, 1],$$

and $\|[\mathcal{F}'(x_0)]^{-1}\mathcal{F}(x_0)\| \leq \|[\mathcal{F}'(x_0)]^{-1}\| \|\mathcal{F}(x_0)\|$. Since

$$[\mathcal{F}(x_0)](s) = -\frac{s}{4} \int_0^1 \frac{dt}{s+t} = -\frac{s}{4} \log \frac{1+s}{s},$$

we have $\|\mathcal{F}(x_0)\| = \frac{\log 2}{4}$. Since

$$\|[I - \mathcal{F}'(x_0)]y\| = \frac{1}{4} \max_{s \in [0,1]} \left| s \int_0^1 \frac{y(t)}{s+t} \, dt + sy(s) \int_0^1 \frac{dt}{s+t} \right| \leq \frac{\log 2}{2} \|y\|,$$

$$\|I - \mathcal{F}'(x_0)\| \leq \frac{\log 2}{2} = 0.3465\ldots < 1,$$

the Banach lemma on invertible operators ensures that the operator $[\mathcal{F}'(x_0)]^{-1}$ exists and

$$\|[\mathcal{F}'(x_0)]^{-1}\| \leq \frac{1}{1 - \|I - \mathcal{F}'(x_0)\|} \leq \frac{1}{1 - \left(\frac{\log 2}{2}\right)} = \frac{2}{\log 2 - 2} = 1.5303\ldots = \beta.$$

Therefore,

$$\|[\mathcal{F}'(x_0)]^{-1}\mathcal{F}(x_0)\| \leq \frac{\left(\frac{\log 2}{4}\right)}{1 - \left(\frac{\log 2}{2}\right)} = \frac{\log 2}{4 - \log 4} = 0.2651\ldots = \eta.$$

Moreover,

$$\|\mathcal{F}'(x) - \mathcal{F}'(y)\| \leq \frac{\log 2}{2} \|x - y\|$$

and so in the present case $L = \frac{\log 2}{2} = 0.3465\ldots$ Furthermore, $a_0 = L\beta\eta = 0.1406\ldots < 1/2$, hence the hypotheses of Theorem 2.1 are satisfied. As a result, Eq. (2.10) with $\lambda = \frac{1}{4}$ has a solution in the ball $\overline{B(x_0, 0.2888\ldots)}$ and the solution is unique in $B(x_0, 1.8853\ldots)$.

After that, we approximate numerically a solution of Eq. (2.10) with $\lambda = \frac{1}{4}$ by Newton's method. For this, we use a discretization process, similar to that used in Sect. 1.2.2, to transform Eq. (2.10) with $\lambda = \frac{1}{4}$ into a finite-dimensional problem. The resulting nonlinear system, whose unknowns are the approximated values of the solution in a series of points of the interval $[0, 1]$, is solved by Newton's method. In addition, we interpolate the discrete solution and obtain an approximation of the solution.

So, we use the Gauss-Legendre formula

$$\int_0^1 \varphi(t)\, dt \simeq \sum_{j=1}^8 w_j \varphi(t_j),$$

where the nodes t_j and the weights w_j are known. If we now denote the approximation of $x(t_i)$ by x_i, for $i = 1, 2, \ldots, 8$, Eq. (2.10) with $\lambda = 1/4$ is equivalent to the nonlinear system of equations

$$x_i = 1 + \frac{1}{4} x_i t_i \sum_{j=1}^8 w_j \frac{x_j}{t_i + t_j}, \quad i = 1, 2, \ldots, 8,$$

which in turn can be written compactly in matrix form:

$$\mathbb{F}(\mathbf{x}) \equiv \mathbf{x} - \mathbf{1} - \frac{1}{4} \mathbf{x} \odot A\mathbf{x} = 0, \qquad \mathbb{F} : \mathbb{R}^8 \longrightarrow \mathbb{R}^8, \qquad (2.11)$$

where $\mathbf{x} = (x_1, x_2, \ldots, x_8)^T$, $\mathbf{1} = (1, 1, \ldots, 1)^T$, $A = (\alpha_{ij})_{i,j=1}^8$, $\alpha_{ij} = \frac{t_i w_j}{t_i + t_j}$ and $\odot$ denotes the inner product.

Since we took $x_0(s) = 1$ as starting point for the theoretical study, a reasonable choice of initial approximation for Newton's method seems to be the vector $\mathbf{x}_0 = \mathbf{1}$. We then first see that conditions of Theorem 2.1 are also satisfied in the discrete case. So, from (2.11), we obtain that the first Fréchet derivative of the function $\mathbb{F}$ is given by

$$\mathbb{F}'(\mathbf{x}) = I - \frac{1}{4} \left(\mathbf{x} \odot (AI) + I \odot (A\mathbf{x}) \right),$$

where I denotes the identity matrix. Consequently,

$$\|\mathbb{F}'(\mathbf{x}) - \mathbb{F}'(\mathbf{y})\| \le L \|\mathbf{x} - \mathbf{y}\| \quad \text{with} \quad L = \frac{1}{2} \|A\| = 0.3446\ldots$$

Moreover,

$$\|[\mathbb{F}'(\mathbf{x}_0)]^{-1}\| \le 1.4806\ldots = \beta, \qquad \|[\mathbb{F}'(\mathbf{x}_0)]^{-1} \mathbb{F}(\mathbf{x}_0)\| \le 0.2403\ldots = \eta,$$

so that $a_0 = L\beta\eta = 0.1226\ldots < 1/2$ and the conditions of Theorem 2.1 are satisfied. Therefore, Newton's method converges to the solution $\mathbf{x}^* = (x_1^*, x_2^*, \ldots, x_8^*)^T$ shown in Table 2.1 after four iterations.

Table 2.1 Numerical solution $\mathbf{x}^*$ of system (2.11)

i	x_i^*	i	x_i^*
1	1.02171973...	5	1.20307175...
2	1.07318638...	6	1.22649087...
3	1.12572489...	7	1.24152460...
4	1.16975331...	8	1.24944851...

Table 2.2 Absolute errors and $\{\|\mathbb{F}(\mathbf{x}_n)\|\}$

n	$\|\mathbf{x}^* - \mathbf{x}_n\|$	$\|\mathbb{F}(\mathbf{x}_n)\|$
0	$2.4944\ldots \times 10^{-1}$	$1.7231\ldots \times 10^{-1}$
1	$9.1396\ldots \times 10^{-3}$	$6.3884\ldots \times 10^{-3}$
2	$9.4963\ldots \times 10^{-6}$	$6.8145\ldots \times 10^{-6}$
3	$8.1882\ldots \times 10^{-12}$	$5.9737\ldots \times 10^{-12}$

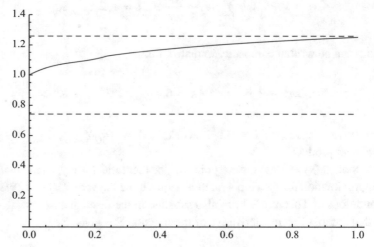

Fig. 2.1 Graph (solid line) of the approximate solution $\mathbf{x}^*$ of system (2.11)

In Table 2.2 we show the errors $\|\mathbf{x}^* - \mathbf{x}_n\|$ and the sequence $\{\|\mathbb{F}(\mathbf{x}_n)\|\}$. Examining this sequence, we see that the vector shown in Table 2.1 is a good approximation of the solution of system (2.11).

Finally, by interpolating the values in Table 2.1 and taking into account that a solution of (2.10) satisfies $x(0) = 1$, we obtain the solution drawn in Fig. 2.1. Observe also that this interpolated solution lies the domain of existence domain of solutions, $\overline{B(\mathbf{1}, 0.2584\ldots)}$, provided by Theorem 2.1.

Operators with Hölder Continuous First Derivative

3

We already explained the difficulties one runs into when we try to use majorizing sequences to prove the semilocal convergence of Newton's method, when we want to relax condition (A2) on F',

$$\|F'(x) - F'(y)\| \le L\|x - y\|, \quad x, y \in \Omega.$$

In this chapter, we extend the technique of recurrence relations presented in Chap. 2, which is an alternative to the technique of majorizing sequences, to prove the semilocal convergence of Newton's method under a milder condition on F' than condition (A2).

We start by assuming that F' is Hölder continuous in Ω [58, 72]:

$$\|F'(x) - F'(y)\| \le K\|x - y\|^p, \quad p \in [0, 1], \quad x, y \in \Omega. \tag{3.1}$$

If $p = 1$, condition (3.1) reduces to condition (A2) and F' is then Lipschitz continuous in Ω.

In this chapter, a semilocal convergence result is given under condition (3.1). Domains of existence and uniqueness of solution of equation $F(x) = 0$ are also provided, the R-order of convergence of Newton's method is studied and a priori error bounds are given.

J. Rokne analyzed the semilocal convergence of Newton's method under condition (3.1), see [72]. However, he did not provide the domain of uniqueness of solution, and the R-order of convergence.

H. B. Keller has also obtained a semilocal convergence result for Newton's method under condition (3.1), see [58], but he did not obtain the R-order of convergence of the method nor the domain of uniqueness of solution. Comparing Keller's result and the one

© The Editor(s) (if applicable) and The Author(s), under exclusive licence to Springer Nature Switzerland AG 2020
J. A. Ezquerro Fernández, M. Á. Hernández Verón, *Mild Differentiability Conditions for Newton's Method in Banach Spaces*, Frontiers in Mathematics,
https://doi.org/10.1007/978-3-030-48702-7_3

presented here, we see that the convergence conditions required in this chapter are milder than those of Keller for certain values of $p \in [0, 1]$.

We then apply the semilocal convergence result to nonlinear Hammerstein integral equations of the second kind and obtain a result on the existence and uniqueness of a solution. The solution of a particular Hammerstein integral equation is then approximated by a discretization process.

3.1 Convergence Analysis

First, under certain conditions on the pair (F, x_0), we study the convergence of Newton's method to a unique solution of the equation $F(x) = 0$ employing the technique based on recurrence relations and introduced in Chap. 2, but in this case under condition (3.1). From some real parameters, a system of two recurrence relations is constructed in which a sequence of positive real numbers is involved and, as a consequence, the convergence of Newton's sequence in the Banach space X is guaranteed.

3.1.1 Recurrence Relations

We assume that the following two conditions:

(H1) There exists $\Gamma_0 = [F'(x_0)]^{-1} \in \mathcal{L}(Y, X)$, for some $x_0 \in \Omega$, with $\|\Gamma_0\| \leq \beta$ and $\|\Gamma_0 F(x_0)\| \leq \eta$;

(H2) $\|F'(x) - F'(y)\| \leq K\|x - y\|^p$, $\quad p \in (0, 1]$, $\quad x, y \in \Omega$.

Denote $a_0 = K\beta\eta^p$ and define the scalar sequence

$$a_{n+1} = a_n f(a_n)^{1+p} g(a_n)^p, \quad n \geq 0, \tag{3.2}$$

where

$$f(t) = \frac{1}{1-t} \quad \text{and} \quad g(t) = \frac{t}{1+p}. \tag{3.3}$$

In addition, we consider $a_0 > 0$, since the problem is trivial if $a_0 = 0$.

Next, we suppose

$$x_1 \in \Omega, \quad a_0 < 1 \tag{3.4}$$

and prove the following two recurrence relations (inequalities) for the sequences (3.2) and $\{x_n\}$:

$$\|\Gamma_1\| = \|[F'(x_1)]^{-1}\| \leq f(a_0)\|\Gamma_0\|, \tag{3.5}$$

$$\|x_2 - x_1\| \leq f(a_0)g(a_0)\|x_1 - x_0\|. \tag{3.6}$$

Since $x_1 \in \Omega$, we have

$$\|I - \Gamma_0 F'(x_1)\| \leq \|\Gamma_0\|\|F'(x_0) - F'(x_1)\| \leq \beta K\|x_1 - x_0\|^p \leq K\beta\eta^p = a_0 < 1$$

and, by the Banach lemma on invertible operators, the operator Γ_1 exists and

$$\|\Gamma_1\| \leq \frac{\|\Gamma_0\|}{1 - \|I - \Gamma_0 F'(x_1)\|} \leq f(a_0)\|\Gamma_0\|.$$

Now, using a Taylor series expansion and Newton's sequence $\{x_n\}$, we obtain

$$\|F(x_1)\| = \left\|\int_0^1 (F'(x_0 + \tau(x_1 - x_0)) - F'(x_0))(x_1 - x_0)\,d\tau\right\| \leq \frac{K\eta^p}{1 + p}\|x_1 - x_0\|$$

and so

$$\|x_2 - x_1\| \leq \|\Gamma_1\|\|F(x_1)\| \leq f(a_0)g(a_0)\|x_1 - x_0\|.$$

Moreover, if $B(x_0, R) \subset \Omega$ and $f(a_0)g(a_0) < 1$, then

$$\|x_2 - x_0\| \leq \|x_2 - x_1\| + \|x_1 - x_0\|$$
$$\leq (1 + f(a_0)g(a_0))\|x_1 - x_0\|$$
$$< \frac{\eta}{1 - f(a_0)g(a_0)}$$
$$= \frac{(1 + p)(1 - a_0)}{(1 + p) - (2 + p)a_0}\eta$$
$$= R \tag{3.7}$$

and $x_2 \in \Omega$.

3.1.2 Analysis of the Scalar Sequence

Here our aim is to analyze the scalar sequence (3.2) and verify that the convergence of Newton's sequence $\{x_n\}$ is guaranteed. To this end, it suffices to show that $\{x_n\}$ is a Cauchy sequence and (3.4) holds for all x_n and a_{n-1} with $n \geq 2$. First, we give a technical lemma whose proof is trivial.

Lemma 3.1 *Let f and g be the two real functions in (3.3). Then*

(a) *f is increasing and $f(t) > 1$ in $(0, 1)$;*
(b) *g is increasing;*
(c) *for $\gamma \in (0, 1)$, we have $f(\gamma t) < f(t)$ if $t \in [0, 1)$ and $g(\gamma t) = \gamma g(t)$.*

Next, we prove several properties of the scalar sequence (3.2). Consider the auxiliary function

$$\phi(t; p) = (1 + p)^p (1 - t)^{1+p} - t^p, \quad p \in (0, 1]. \tag{3.8}$$

This function has only one zero $\xi(p)$ in the interval $(0, 1/2]$, since $\phi(0; p) = (1+p)^p > 0$, $\phi(1/2; p) \leq 0$ and $\phi'(t; p)(t) < 0$ in $(0, 1/2]$. Notice that the function (3.8) arises by setting the value $f(a_0)^{1+p} g(a_0)^p = \frac{a_0^p}{(1+p)^p(1-a_0)^{1+p}}$ equal to 1.

Lemma 3.2 *Let f and g be the two scalar functions in (3.3). If $a_0 \in (0, \xi(p))$, where $\xi(p)$ is the unique zero of the function (3.8) in the interval $(0, 1/2]$ and $p \in (0, 1]$, then*

(a) *$f(a_0)^{1+p} g(a_0)^p < 1$;*
(b) *the sequence $\{a_n\}$ is strictly decreasing;*
(c) *$a_n < 1$, for all $n \geq 0$.*

If $a_0 = \xi(p)$, then $a_n = a_0 < 1$ for all $n \geq 1$.

Proof Suppose first that $a_0 \in (0, \xi(p))$. Then, item (a) follows from the fact that $\phi(a_0; p) > 0$. Next, item (b) is proved by induction on n. As $f(a_0)^{1+p} g(a_0)^p < 1$, we have $a_1 < a_0$. If we now assume that $a_j < a_{j-1}$, for $j = 1, 2, \ldots, n$, then

$$a_{n+1} = a_n f(a_n)^{1+p} g(a_n)^p < a_n f(a_0)^{1+p} g(a_0)^p < a_n,$$

since f and g are increasing. Thus, the sequence $\{a_n\}$ is strictly decreasing. To see item (c), note that $a_n < a_0 < 1$, for all $n \geq 0$, from the fact that the sequence $\{a_n\}$ is strictly decreasing and $a_0 \in (0, \xi(p))$.

Finally, if $a_0 = \xi(p)$, then $f(a_0)^{1+p}g(a_0)^p = 1$ and $a_n = a_0 = \xi(p) < 1$, for all $n \geq 0$. ∎

Lemma 3.3 *Let f and g be the two real functions given in* (3.3). *If $a_0 \in (0, \xi(p))$, where $\xi(p)$ is the unique zero of the function* (3.8) *in the interval $(0, 1/2]$ and $p \in (0, 1]$, set $\gamma = \frac{a_1}{a_0}$. Then*

(a) $a_n < \gamma^{(1+p)^{n-1}} a_{n-1}$ *and* $a_n < \gamma^{\frac{(1+p)^{n-1}-1}{p}} a_0$, *for all $n \geq 2$;*

(b) $f(a_n)g(a_n) < \gamma^{\frac{(1+p)^{n-1}-1}{p}} f(a_0)g(a_0) = \dfrac{\gamma^{\frac{(1+p)^n}{p}}}{f(a_0)^{1/p}}$, *for all $n \geq 1$.*

If $a_0 = \xi(p)$, then $f(a_n)g(a_n) = f(a_0)g(a_0) = \frac{1}{f(a_0)^{1/p}}$, for all $n \geq 1$.

Proof We only prove the case $a_0 \in (0, \xi(p))$, since proof for the case $a_0 = \xi(p)$ is similar. The proof of item (a) is carried out by induction on n. If $n = 2$, then by Lemma 3.1 (b),

$$a_2 = a_1 f(a_1)^{1+p}g(a_1)^p = \gamma a_0 f(\gamma a_0)^{1+p}g(\gamma a_0)^p < \gamma^{1+p}a_1 = \gamma^{2+p}a_0.$$

Now suppose that

$$a_{n-1} < \gamma^{(1+p)^{n-2}} a_{n-2} < \gamma^{\frac{(1+p)^{n-1}-1}{p}} a_0.$$

Then, proceeding in the same way as for $n = 2$, we get that

$$\begin{aligned}
a_n &= a_{n-1}f(a_{n-1})^{1+p}g(a_{n-1})^p \\
&< \gamma^{(1+p)^{n-2}} a_{n-2} f\left(\gamma^{(1+p)^{n-2}} a_{n-2}\right)^{1+p} g\left(\gamma^{(1+p)^{n-2}} a_{n-2}\right)^p \\
&< \gamma^{(1+p)^{n-1}} a_{n-1} \\
&< \gamma^{(1+p)^{n-1}} \gamma^{(1+p)^{n-2}} a_{n-2} \\
&< \cdots < \gamma^{\frac{(1+p)^n-1}{p}} a_0.
\end{aligned}$$

To prove item (b), we observe that, for $n \geq 1$,

$$f(a_n)g(a_n) < f\left(\gamma^{\frac{(1+p)^{n-1}-1}{p}} a_0\right) g\left(\gamma^{\frac{(1+p)^{n-1}-1}{p}} a_0\right) < \gamma^{\frac{(1+p)^{n-1}-1}{p}} f(a_0)g(a_0) = \dfrac{\gamma^{\frac{(1+p)^n}{p}}}{f(a_0)^{1/p}}.$$

The proof is complete. ∎

3.1.3 A Semilocal Convergence Result

We are now ready to prove the semilocal convergence of Newton's method when it is applied to operators with Hölder continuous first Fréchet derivative. In addition, we introduce the condition

(H3) $a_0 = K\beta\eta^p \in (0, \xi(p)]$, where $\xi(p)$ is the unique zero of function (3.8) in the interval $(0, 1/2]$, $p \in (0, 1]$ and $B(x_0, R) \subset \Omega$, where $R = \frac{(1+p)(1-a_0)}{(1+p)-(2+p)a_0}\eta$.

Theorem 3.4 *Let $F : \Omega \subseteq X \to Y$ be a continuously Fréchet differentiable operator defined on a nonempty open convex domain Ω of a Banach space X with values in a Banach space Y. Suppose that conditions (H1)-(H2)-(H3) are satisfied. Then, the Newton sequence $\{x_n\}$ starting at x_0 converges to a solution x^* of $F(x) = 0$. Moreover, $x_n, x^* \in \overline{B(x_0, R)}$ and x^* is unique in $B\left(x_0, \eta/a_0^{1/p}\right) \cap \Omega$. The Newton sequence $\{x_n\}$ has R-order of convergence at least $1 + p$ if $a_0 \in (0, \xi(p))$, or at least 1 if $a_0 = \xi(p)$, and*

$$\|x^* - x_n\| \leq \left(\gamma^{\frac{(1+p)^n - 1}{p^2}}\right) \frac{\Delta^n}{1 - \gamma^{\frac{(1+p)^n}{p}}\Delta}\eta, \quad n \geq 0, \tag{3.9}$$

where $\gamma = \frac{a_1}{a_0}$ and $\Delta = (1 - a_0)^{1/p}$.

Proof We start with the case $a_0 \in (0, \xi(p))$. First, we prove that the Newton sequence $\{x_n\}$ satisfies the following items for $n \geq 2$:

(I$_n$)   There exists $\Gamma_{n-1} = [F'(x_{n-1})]^{-1}$ and $\|\Gamma_{n-1}\| \leq f(a_{n-2})\|\Gamma_{n-2}\|$,

(II$_n$) $\|x_n - x_{n-1}\| \leq f(a_{n-2})g(a_{n-2})\|x_{n-1} - x_{n-2}\|$,

(III$_n$) $x_n \in \Omega$.

Observe that $x_1 \in \Omega$, since $\eta < R$. Then, (3.5), (3.6) imply that the items above hold for $n = 2$. If we now suppose that items (I$_{n-1}$)-(II$_{n-1}$) are true, it follows, by analogy to case $n = 2$, that items (I$_n$)-(II$_n$) also hold.

Notice that $a_n < 1$ for all $n \geq 0$. Let us prove item (III$_n$). Item (II$_n$) and Lemma 3.3 (b) imply that

$$\|x_n - x_0\| \leq \|x_n - x_{n-1}\| + \|x_{n-1} - x_{n-2}\| + \cdots + \|x_1 - x_0\|$$

$$\leq \left(1 + \sum_{i=0}^{n-2}\left(\prod_{j=0}^{i} f(a_j)g(a_j)\right)\right)\|x_1 - x_0\|$$

$$< \left(1 + \sum_{i=0}^{n-2}\left(\prod_{j=0}^{i} f(a_0)g(a_0)\gamma^{\frac{(1+p)^j - 1}{p}}\right)\right)\|x_1 - x_0\|$$

$$= \left(1 + \sum_{i=0}^{n-2}\left(\prod_{j=0}^{i}\left(\gamma^{\frac{(1+p)^j}{p}}\Delta\right)\right)\right)\|x_1 - x_0\|$$

$$= \left(1 + \sum_{i=0}^{n-2}\left(\gamma^{\frac{(1+p)^{1+i}-1}{p^2}}\Delta^{1+i}\right)\right)\|x_1 - x_0\|,$$

where $\gamma = \frac{a_1}{a_0} < 1$ and $\Delta = \frac{f(a_0)g(a_0)}{\gamma^{1/p}} = \frac{1}{f(a_0)^{1/p}} = (1 - a_0)^{1/p} < 1$. By the Bernoulli inequality,

$$\gamma^{\frac{(1+p)^{1+i}-1}{p^2}} = \gamma^{1/p}\gamma^{\frac{1+p}{p^2}((1+p)^i-1)} \leq \gamma^{1/p}\gamma^{\frac{1+p}{p}i}.$$

Consequently,

$$\|x_n - x_0\| < \left(1 + \gamma^{1/p}\Delta\sum_{i=0}^{n-2}\gamma^{\frac{1+p}{p}i}\Delta^i\right)\|x_1 - x_0\|$$

$$< \left(1 + \gamma^{1/p}\Delta\frac{1 - \left(\gamma^{\frac{1+p}{p}}\Delta\right)^{n-1}}{1 - \gamma^{\frac{1+p}{p}}\Delta}\right)\eta$$

$$< \frac{\eta}{1 - \gamma^{1/p}\Delta}$$

$$= \frac{(1 + p)(1 - a_0)}{(1 + p) - (2 + p)a_0}\eta$$

$$= R$$

and $x_n \in B(x_0, R)$. As $B(x_0, R) \subset \Omega$, the approximation $x_n \in \Omega$, for all $n \geq 0$. Note that the conditions required in (3.4) are now satisfied for all x_n and a_{n-1} with $n \geq 2$.

Second, we prove that $\{x_n\}$ is a Cauchy sequence. For this, we proceed in much the same way as before. So, for $m \geq 1$ and $n \geq 1$, item (II_n) and Lemma 3.3 (h) imply that

$$\|x_{n+m} - x_n\| \leq \|x_{n+m} - x_{n+m-1}\| + \|x_{n+m-1} - x_{n+m-2}\| + \cdots + \|x_{n+1} - x_n\|$$

$$\leq \sum_{i=n-1}^{n+m-2}\left(\prod_{j=0}^{i}f(a_j)g(a_j)\right)\|x_1 - x_0\|$$

$$< \sum_{i=n-1}^{n+m-2}\left(\prod_{j=0}^{i}\gamma^{\frac{(1+p)^j-1}{p}}f(a_0)g(a_0)\right)\|x_1 - x_0\|$$

$$= \sum_{i=n-1}^{n+m-2} \left(\prod_{j=0}^{i} \left(\gamma^{\frac{(1+p)^j}{p}} \Delta \right) \right) \|x_1 - x_0\|$$

$$= \sum_{i=n-1}^{n+m-2} \left(\gamma^{\frac{(1+p)^{1+i}-1}{p^2}} \Delta^{1+i} \right) \|x_1 - x_0\|$$

$$= \sum_{i=0}^{m-1} \left(\gamma^{\frac{(1+p)^{n+i}-1}{p^2}} \Delta^{n+i} \right) \|x_1 - x_0\|.$$

By the Bernoulli inequality,

$$\gamma^{\frac{(1+p)^{n+i}-1}{p^2}} = \gamma^{\frac{(1+p)^n-1}{p^2}} \gamma^{\frac{(1+p)^n}{p^2}((1+p)^i-1)} \leq \gamma^{\frac{(1+p)^n-1}{p^2}} \gamma^{\frac{(1+p)^n}{p}i},$$

so that

$$\|x_{n+m} - x_n\| < \left(\sum_{i=0}^{m-1} \left(\gamma^{\frac{(1+p)^n}{p}i} \Delta^i \right) \right) \gamma^{\frac{(1+p)^n-1}{p^2}} \Delta^n \|x_1 - x_0\|$$

$$< \frac{1 - \left(\gamma^{\frac{(1+p)^n}{p}} \Delta \right)^m}{1 - \gamma^{\frac{(1+p)^n}{p}} \Delta} \gamma^{\frac{(1+p)^n-1}{p^2}} \Delta^n \eta. \tag{3.10}$$

Therefore, $\{x_n\}$ is a Cauchy sequence, as claimed, and consequently there exists $x^* \in \overline{B(x_0, R)}$ such that $x^* = \lim_n x_n$.

Third, we prove that x^* is a solution of the equation $F(x) = 0$. As $\|\Gamma_n F(x_n)\| \to 0$ when $n \to \infty$, if we take into account that

$$\|F(x_n)\| \leq \|F'(x_n)\| \|\Gamma_n F(x_n)\|$$

and $\{\|F'(x_n)\|\}$ is bounded, since

$$\|F'(x_n)\| \leq \|F'(x_0)\| + K\|x_n - x_0\|^p < \|F'(x_0)\| + KR^p,$$

we see that $\|F(x_n)\| \to 0$ when $n \to \infty$. In addition, $F(x^*) = 0$ thanks to the continuity of F in $\overline{B(x_0, R)}$.

Fourth, to prove the uniqueness of the solution x^*, we assume that y^* is another solution of $F(x) = 0$ in $B\left(x_0, \eta/a_0^{1/p}\right) \cap \Omega$. Then, from the approximation

$$0 = F(y^*) - F(x^*) = \int_{x^*}^{y^*} F'(x)dx = \int_0^1 F'(x^* + \tau(y^* - x^*)) \, d\tau(y^* - x^*),$$

it follows that $x^* = y^*$, provided that the operator $\int_0^1 F'(x^* + \tau(y^* - x^*)) \, d\tau$ is invertible. To verify that this is the case, we prove equivalently that the operator $J = \Gamma_0 \int_0^1 F'(x^* + \tau(y^* - x^*)) \, d\tau$ is invertible. Indeed, as

$$\|I - J\| \leq \|\Gamma_0\| \int_0^1 \|F'(x^* + \tau(y^* - x^*)) - F'(x_0)\| \, d\tau$$

$$\leq K\beta \int_0^1 \|x^* + \tau(y^* - x^*) - x_0\|^p \, d\tau$$

$$\leq K\beta \int_0^1 ((1 - \tau)\|x^* - x_0\| + \tau\|y^* - x_0\|)^p \, d\tau$$

$$\leq K\beta \int_0^1 \left(R(1 - \tau) + \left(\frac{\eta}{a_0^{1/p}} \right) \tau \right)^p \, d\tau$$

$$< K\beta \int_0^1 \left(\frac{\eta}{a_0^{1/p}} \right)^p \, d\tau$$

$$= 1,$$

the operator J^{-1} exists by the Banach lemma on invertible operators.

Finally, letting $m \to \infty$ in (3.10), we obtain (3.9) for all $n \geq 0$. Moreover, from (3.9), it follows that the R-order of convergence of the sequence $\{x_n\}$ is at least $1 + p$, since

$$\|x^* - x_n\| \leq \frac{\eta}{\gamma^{1/p^2}(1 - \gamma^{1/p}\Delta)} \left(\gamma^{1/p^2} \right)^{(1+p)^n}, \quad n \geq 0.$$

For the second case, $a_0 = \xi(p)$, we have $a_n = a_0 = \xi(p)$, for all $n \geq 0$. Then, following arguing similarly to the previous case and taking into account that $\gamma = 1$ and $\Delta = f(a_0)g(a_0) < 1$, we obtain the same results except for the R-order of convergence, which is at least one in this case. ∎

As in all the results of semilocal convergence (see, for example, [2, 7, 16, 17]), we observe in Theorem 3.4 that there are conditions on the operator involved (condition (H2)), on the starting point (condition (H1)) and on the relationship between both (condition (H3)).

3.2   A Little More

In this section, make several imporant comments. First, we provide a semilocal convergence result for Newton's method if $p = 0$ in (3.1), which was not included in Sect. 3.1. Second, we discuss the case $p = 1$ in (3.1). Third, we give another analysis of the scalar sequence (3.2), different from the one in Sect. 3.1.2, that leads to the same results. Finally, we present a study about the starting points and compare the conditions required in Theorem 3.4 with those required in [58] by Keller.

3.2.1   Case $p = 0$

If $p = 0$, conditions (H2)–(H3) are reduced respectively to

$(\widetilde{H2})$ $\quad \|F'(x) - F'(y)\| \leq K, \quad x, y \in \Omega, \quad K \geq 0.$

$(\widetilde{H3})$ $\quad a_0 = K\beta \in \left(0, \frac{1}{2}\right)$ and and $B(x_0, r) \subset \Omega$, where $R = \frac{1-a_0}{1-2a_0}\eta.$

Arguing similarly to the case $p \in (0, 1]$, we see that

- $\|\Gamma_n\| \leq \dfrac{\|\Gamma_0\|}{1 - a_0}$, for all $n \geq 1$, if $a_0 < 1$ and $x_n \in \Omega$,

- $\|x_{n+1} - x_n\| \leq \left(\dfrac{a_0}{1 - a_0}\right)^n \|x_1 - x_0\|, \ n \geq 0.$

This yields the following semilocal convergence result for Newton's method.

Theorem 3.5 *Let X and Y be two Banach spaces and $F : \Omega \subseteq X \to Y$ a continuously Fréchet differentiable operator on a nonempty open convex domain Ω. Suppose that conditions (H1)-$(\widetilde{H2})$-$(\widetilde{H3})$ are satisfied. Then, the Newton sequence $\{x_n\}$ starting at x_0 converges to a solution x^* of the equation $F(x) = 0$. Moreover, $x_n, x^* \in \overline{B(x_0, r)}$ and x^* is unique in Ω.*

3.2.2   Case $p = 1$

If $p = 1$ in the study presented in Sect. 3.1, Theorem 3.4 reduces to Theorem 2.1. In addition, inequalities (I_n)–(II_n), which were used in the proof of Theorem 3.4 reduce to equalities when the equation to solve is

$$\psi(t) = \frac{K}{2}t^2 - \frac{t}{\beta} + \frac{\eta}{\beta} = 0.$$

Hence, (I_n)–(II_n) are optimal in this case and take the form of equalities. Thanks to this, we could improve the a priori error bounds given by other authors. Notice that the polynomial ψ is just the quadratic polynomial of Kantorovich that is ordinarilly used to construct a majorizing sequence and prove the semilocal convergence of Newton's method if condition (K3) is replaced by condition (A2), see [62].

3.2.3 Scalar Sequence

We can also analyze the scalar sequence (3.2) by studying the fixed points of the function $\varphi(t) = t f(t)^{1+p} g(t)^p$. If $\varphi(t) = t$, then t is a fixed point of φ, i.e., if $t \neq 0, 1$, then

$$\frac{t^{1+p}}{(1+p)^p (1-t)^{1+p}} = t \qquad \Longleftrightarrow \qquad (1+p)^p (1-t)^{1+p} - t^p = 0,$$

which agrees with $\phi(t; p) = 0$, where ϕ is defined in (3.8).

As we can see in Fig. 3.1, $t = 0$ and $t = 1/2$ are fixed points of φ in $[0, 1]$ if $p = 1$. It is then easy to prove that the sequence $\{a_n = \varphi(a_{n-1})\}$ only converges if $a_0 \leq 1/2$, so that

$$a_n \searrow 0 \quad (n \to \infty) \text{ if } a_0 < 1/2,$$
$$a_n = 1/2 \ (n \geq 0) \quad \text{if } a_0 = 1/2.$$

Both situations appear in Lemma 3.2.

On the other hand, if the value $p \in (0, 1)$ varies, the fixed points that appear are $t = 0$ and $t = \xi(p)$ (zero of ϕ) with $\xi(p) < 1/2$, see Fig. 3.2.

In both cases, it is interesting to observe that, if $a_0 > \xi(p) = 1/2 \ (p = 1)$ or $a_0 > \xi(p)$ $(0 < p < 1)$, then $n_0 \in \mathbb{N}$ exists such that $a_{n_0} > 1$, see Fig. 3.3, so that the condition $a_n < 1$, for all $n \geq 0$, is not satisfied.

Fig. 3.1 Fixed points of φ when $p = 1$

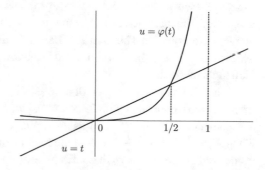

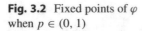

Fig. 3.2 Fixed points of φ when $p \in (0, 1)$

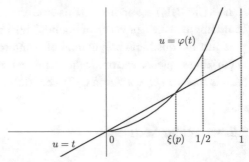

Fig. 3.3 The case when $a_0 > \xi(p)$

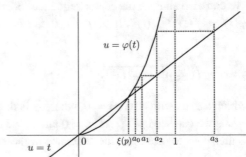

3.2.4   Starting Points

As we have indicated in Sect. 1.3, the basins of attraction do not depend on the conditions required in the result established for the semilocal convergence of an iterative method; rather, they depend only on the equation to be solved, so that we do not analyze them further. We focus our attention on the regions of accessibility and the domains of parameters, both of which depend on the semilocal convergence results.

3.2.4.1 Region of Accessibility
We have seen in Sect. 1.3.2 that with the starting point of Newton's method there are associated the parameters L, β and η of conditions (A1)–(A2)–(A3) imposed in the Newton-Kantorovich Theorem 2.1. Based on these conditions, we present the regions of accessibility of the three solutions of equation $z^3 - 1 = 0$ when it is approximated by Newton's method.

Figure 3.4 shows the region of accessibility of the unique solution $z^* = 1$ of equation $z^{3/2} - 1 = 0$ associated with Theorem 3.4 when it is approximated by Newton's method. We consider the ball $B(0, 1.2)$ as the domain of the complex function $d(z) = z^{3/2} - 1$. Then $K = \frac{3}{2}$ and $p = \frac{1}{2}$. To paint the regions of accessibility, we take $z_0 \in B(0, 1.2)$ and colour red all the points z_0 that satisfy condition (H3) of Theorem 3.4: $K\beta\eta^p \in (0, \xi(p)]$, where $\xi(p) = \xi\left(\frac{1}{2}\right) = 0.3718\ldots$

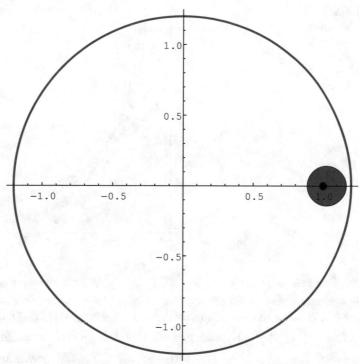

Fig. 3.4 Region of accessibility of the solution $z^* = 1$ of equation $z^{3/2} - 1 = 0$ associated with Theorem 3.4

3.2.4.2 Domain of Parameters

Next, we picture the domain of parameters of Newton's method associated with Theorem 3.4, which is given by

$$D = \left\{ (x, y) \in \mathbb{R}_+^2 : xy \leq \xi(p) \right\},$$

where we now choose $x = \beta$ and $y = K\eta^p$, so that $K\beta\eta^p \in (0, \xi(p)]$, where $\xi(p)$ is the unique zero of function (3.8) in the interval $(0, 1/2]$, of Theorem 3.4 is satisfied. In particular, in Fig. 3.5, we see the domains of parameters of Newton's method associated with Theorem 3.4 for different values of p. Observe that the larger the value of $p \in (0, 1]$, the bigger the domain of parameters: yellow for $p = \frac{1}{2}$, red for $p = \frac{2}{3}$, green for $p = \frac{5}{6}$ and black for $p = 1$. Note that the domain of parameters obtained for $p = 1$ is the same as that associated with the Newton-Kantorovich Theorem 2.1, see Fig. 1.6 and Sect. 1.3.3.

3.2.4.3 Comparative Study

We compare the conditions required for the convergence of Newton's method in Theorem 3.4 and those appearing in Keller's theorem (Theorem 4 of [58]).

Fig. 3.5 Domains of parameters of Newton's method associated with Theorem 3.4 when $p = \frac{1}{2}, \frac{2}{3}, \frac{5}{6}, 1$ (yellow, red, green and black, respectively)

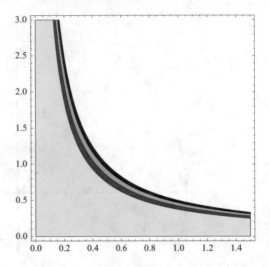

Theorem 3.6 (Theorem 4 of [58]) *Let $F : \Omega \subseteq X \to Y$ be a continuously Fréchet differentiable operator defined on a nonempty open convex domain Ω of a Banach space X with values in a Banach space Y. Suppose that the conditions (H1)-(H2) are satisfied and assume $K\beta\eta^p \leq \frac{1}{2+p}\left(\frac{p}{1+p}\right)^p$, with $p \in (0, 1]$. Let $R_0(p)$ be the smallest positive root of the equation $(2 + p)K\beta t^{1+p} - (1 + p)(t - \eta) = 0$. If $\rho \geq R_0(p)$, then the equation $F(x) = 0$ has a solution $x^* \in \overline{B(x_0, \rho)}$ and the Newton sequence with initial point x_0 converges to this solution.*

Observe that Keller proves the convergence of Newton's method under the same general hypotheses as the Newton-Kantorovich Theorem 2.1, except that he relaxes the requirement that F' is Lipschitz continuous in Ω to F' is Hölder continuous in Ω. Although Theorem 3.6 is an extension of Theorem 2.1, it does not formally reduce to that case when $p = 1$, since Keller obtains $K\beta\eta \leq \frac{1}{6}$ in place of $K\beta\eta \leq \frac{1}{2}$ as in (A3) (recall that here K is the Lispchitz constant). He proves another theorem that reduces to the Newton-Kantorovich theorem for $p = 1$, but it is not valid for $p = 0$. J. Rokne proves in [72] a similar theorem that is valid for $p \in [0, 1]$ and reduces to Keller's theorem for $p = 0$, but he is only able to prove linear convergence. For this, Rokne uses the concept of majorizing sequence introduced by W. C. Rheinboldt in [71] and proves the theorem under the much more general conditions of Rheinboldt (see Theorem 4.3 of [71]), that is, instead of considering the ordinary Newton method, he considers the general form

$$x_{n+1} = x_n - [\mathcal{A}(x_n)]^{-1}F(x_n), \quad n \geq 0, \quad \text{with } x_0 \text{ given,}$$

where $\mathcal{A}(x_n)$ is an approximation to $F'(x_n)$. Finally, we note that the general convergence theory presented by Rheinboldt in [71] reduces, in its simplest form, the study of the iterative methods to that of a second-order difference equation, where the concept of majorizing sequence is based on a simple principle observed to underlie Kantorovich's majorant proof of the convergence of Newton's method.

For each p fixed in $(0, 1]$, the domain of parameters that we can associate with Theorem 3.6 is

$$D_K = \left\{ (x, y) \in \mathbb{R}_+^2 : xy \leq \frac{1}{2+p} \left(\frac{p}{1+p} \right)^p, \; p \in (0, 1] \right\},$$

where $x = \beta$ and $y = K\eta^p$.

Let us compare the conditions required for the semilocal convergence of Newton's method in Theorems 3.4 and 3.6. Observe that

$$\frac{1}{2+p} \left(\frac{p}{1+p} \right)^p < \xi(p) \quad \text{if} \quad p \in (0.2856\ldots, 1],$$

so that $D_K \subset D$ if $p \in (0.2856\ldots, 1]$, as we can see in Fig. 3.6, for $p = \frac{1}{10}, \frac{1}{5}, \frac{2}{5}, \frac{1}{2}, \frac{7}{10}, \frac{4}{5}$. Therefore, the chances of finding starting points in the application of Newton's method for operators with Hölder continuous first Fréchet derivative and $p \in (0.2856\ldots, 1]$ are higher if Theorem 3.4 is applied.

3.3   Application to a Hammerstein Integral Equation

As we have seen, an interesting by product of the study of the convergence of iterative methods for solving equations is than one can obtain results of existence and uniqueness of solution for the equations. Thus, we can use the theoretical power of Newton's method to draw conclusions about the existence and uniqueness of a solution and about the region in which it is located, without finding the solution itself; sometimes this is more important than the actual knowledge of the solution. In particular, using Theorem 3.4 we can locate and separate solutions of Eq. (3.11) below and guarantee the semilocal convergence of Newton's method to a solution of it.

Then, we provide some results of this type for Hammerstein integral equations of type (1.6). In particular, we consider the equation

$$x(s) = u(s) + \lambda \int_a^b \mathcal{K}(s, t) x(t)^{1+p} \, dt, \quad s \in [a, b], \quad p \in [0, 1], \quad \lambda \in \mathbb{R}, \qquad (3.11)$$

where $-\infty < a < b < +\infty$, $u(s)$ is a continuous function in $[a, b]$, the kernel $\mathcal{K}(s, t)$ is a known function in $[a, b] \times [a, b]$, and $x(s)$ is the unknown function. Problems of this type are used as "tests" in the study of operators with Hölder continuous first Fréchet derivative, see [8, 72].

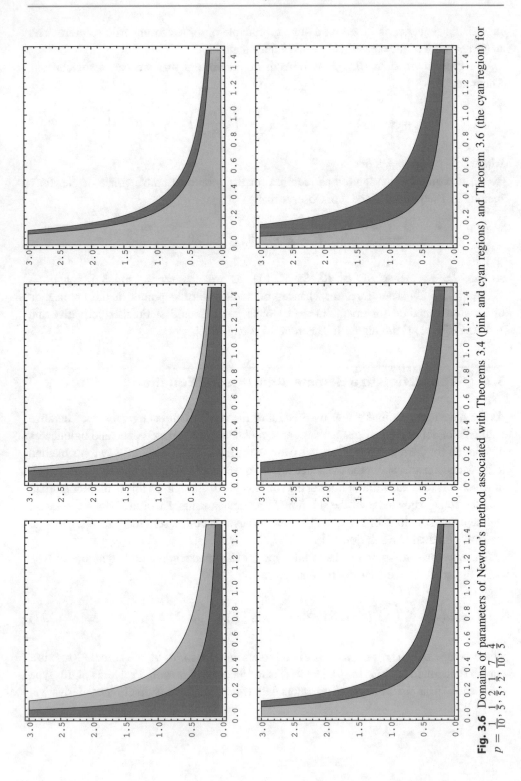

Fig. 3.6 Domains of parameters of Newton's method associated with Theorems 3.4 (pink and cyan regions) and Theorem 3.6 (the cyan region) for $p = \frac{1}{10}, \frac{1}{5}, \frac{2}{5}, \frac{1}{2}, \frac{7}{10}, \frac{4}{5}$

Notice that (3.11) is a modification of the equation

$$x(s) = u(s) + \lambda \int_a^b \mathcal{K}(s, t)x(t)^q \, dt, \quad s \in [a, b], \quad \lambda \in \mathbb{R}, \quad q \in \mathbb{N}, \quad (3.12)$$

which is studied is given in [24].

3.3.1   Existence and Uniqueness of a Solution

Observe that solving (3.11) is equivalent to solving the equation $\mathcal{F}(x) = 0$, where

$$\mathcal{F} : \Omega \subseteq \mathcal{C}([a, b]) \longrightarrow \mathcal{C}([a, b]), \quad \Omega = \{x \in \mathcal{C}([a, b]) : x(s) > 0, \; s \in [a, b]\}, \quad (3.13)$$

$$[\mathcal{F}(x)](s) = x(s) - u(s) - \lambda \int_a^b \mathcal{K}(s, t)x(t)^{1+p} \, dt, \quad p \in [0, 1], \quad \lambda \in \mathbb{R}. \quad (3.14)$$

It is clear that Theorem 3.4 can be used to establish the existence of a solution of Eq. (3.11) in $\overline{B(x_0, R)}$ and its uniqueness in $B\left(x_0, \frac{\eta}{a_0^{1/p}}\right) \cap \Omega$. To this end we need to calculate the parameters K, β and η based on the starting function $x_0(s)$. First, we have

$$[\mathcal{F}'(x)y](s) = y(s) - (1 + p)\lambda \int_a^b \mathcal{K}(s, t)x(t)^p y(t) \, dt.$$

If $x_0(s)$ is fixed,

$$\|I - \mathcal{F}'(x_0)\| \le (1 + p)|\lambda|\ell\|x_0^p\|,$$

where $\ell = \max\limits_{s \in [a,b]} \left|\int_a^b \mathcal{K}(s, t) \, dt\right|$. By the Banach lemma on invertible operators, if $(1 + p)|\lambda|\ell\|x_0^p\| < 1$, then

$$\|[\mathcal{F}'(x_0)]^{-1}\| \le \frac{1}{1 - (1 + p)|\lambda|\ell\|x_0^p\|}.$$

From the definition of the operator $\mathcal{F}$ in (3.13)–(3.14), it follows that $\|\mathcal{F}(x_0)\| \le \|x_0 - u\| + |\lambda|\ell\left\|x_0^{1+p}\right\|$ and

$$\|[\mathcal{F}'(x_0)]^{-1}\mathcal{F}(x_0)\| \le \frac{\|x_0 - u\| + |\lambda|\ell\left\|x_0^{1+p}\right\|}{1 - (1 + p)|\lambda|\ell\left\|x_0^p\right\|}.$$

Furthermore, as

$$(\mathcal{F}'(x) - \mathcal{F}'(y))z(s) = -(1+p)\lambda \int_a^b \mathcal{K}(s,t)\left(x(t)^p - y(t)^p\right)z(t)\,dt$$

and $\left|t_1^p - t_2^p\right| \le |t_1 - t_2|^p$, we find that

$$\|\mathcal{F}'(x) - \mathcal{F}'(y)\| \le (1+p)|\lambda|\ell\|x - y\|^p.$$

Therefore, once the parameters

$$\beta = \frac{1}{1 - (1+p)|\lambda|\ell\|x_0^p\|}, \quad \eta = \frac{\|x_0 - u\| + |\lambda|\ell\left\|x_0^{1+p}\right\|}{1 - (1+p)|\lambda|\ell\|x_0^p\|}, \quad K = (1+p)|\lambda|\ell$$

(3.15)

are determined for Theorem 3.4, the following result of existence and uniqueness of a solution for Eq. (3.11) is obtained.

Theorem 3.7 *Let $\mathcal{F}$ be the operator defined in (3.13)–(3.14) and let $x_0 \in \Omega \subseteq C([a, b])$ be a function such that there exists the operator $[\mathcal{F}'(x_0)]^{-1}$. If*

$$(1+p)|\lambda|\ell\|x_0^p\| < 1, \qquad a_0 = K\beta\eta^p \in (0, \xi(p)]$$

where K, β, η are defined in (3.15), $p \in (0, 1]$ and $\xi(p)$ is the unique zero of the function (3.8) in $(0, 1/2]$, and $B(x_0, R) \subset \Omega$, where $R = \frac{(1+p)(1-a_0)}{(1+p)-(2+p)a_0}\eta$, then Eq. (3.11) admits a solution in $\overline{B(x_0, R)}$. Moreover, this solution is unique in $B\left(x_0, \eta/a_0^{1/p}\right) \cap \Omega$.

By analogy, if $p = 0$, a similar theorem can be proved; see Sect. 3.2.1.

Note that the upper bound given for $\|\mathcal{F}(x_0)\|$ is improved once the kernel $\mathcal{K}$ and the function u are known.

3.3.2 Localization of Solutions

Let us consider the following particular case of Eq. (3.11):

$$x(s) = 1 + \lambda \int_0^1 \mathcal{G}(s, t)x(t)^{3/2}\,dt,$$

(3.16)

where $\mathcal{G}(s, t)$ is the Green function given in (1.7), with $[a, b] = [0, 1]$. Our immediate aim is to obtain a result, depending on the parameter λ, on the existence and uniqueness of a solution for Eq. (3.16). Next, (3.16) is discretized and Newton's method is applied to obtain an approximate solution.

If we proceed as (3.11) with $[a, b] = [0, 1]$, $u(s) = 1$ and $p = 1/2$, we see, by the Banach lemma on invertible operators, that the operator $[\mathcal{F}'(x_0)]^{-1}$ exists if

$$|\lambda| < \frac{16}{3 \left\| x_0^{1/2} \right\|}, \tag{3.17}$$

since $\max\limits_{s \in [0,1]} \left| \int_0^1 \mathcal{G}(s, t) \, dt \right| = \dfrac{1}{8}$ and

$$\| I - \mathcal{F}'(x_0) \| \leq \frac{3}{16} |\lambda| \left\| x_0^{1/2} \right\| < 1.$$

Moreover,

$$\| [\mathcal{F}'(x_0)]^{-1} \| \leq \frac{16}{16 - 3|\lambda| \left\| x_0^{1/2} \right\|}.$$

We also have

$$\| \mathcal{F}(x_0) \| \leq \| x_0 - 1 \| + \frac{|\lambda|}{8} \left\| x_0^{3/2} \right\|$$

and

$$\| \mathcal{F}'(x) - \mathcal{F}'(y) \| \leq \frac{3}{16} |\lambda| \| x - y \|^{1/2}, \quad x, y \in \Omega.$$

Hence, taking into account that

$$\beta = \frac{16}{16 - 3|\lambda| \left\| x_0^{1/2} \right\|}, \qquad \eta = \frac{16 \left(\| x_0 - 1 \| + \frac{|\lambda|}{8} \left\| x_0^{3/2} \right\| \right)}{16 - 3|\lambda| \left\| x_0^{1/2} \right\|}, \qquad K = \frac{3}{16} |\lambda|,$$

and condition (3.17), which is shown in Fig. 3.7, where $\left\| x_0^{1/2} \right\|$ is represented in the x-axis and $|\lambda|$ in the y-axis, we obtain the following corollary.

Corollary 3.8 *Let $\mathcal{F}$ be corresponding operator defined from (3.13)–(3.14) as*

$$[\mathcal{F}(x)](s) = x(s) - 1 - \lambda \int_0^1 \mathcal{G}(s, t) x(t)^{3/2} \, dt.$$

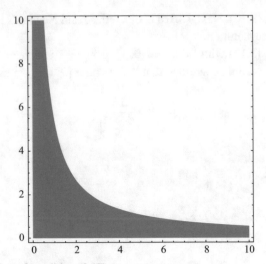

Fig. 3.7 Representation of condition (3.17)

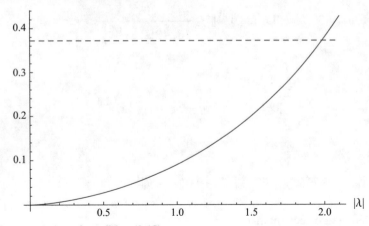

Fig. 3.8 Representation of condition (3.18)

*If $x_0 \in \Omega$, λ satisfies (3.17), $a_0 = K\beta\eta^{1/2} \leq \xi(1/2) = 0.3718\dots$ and $B(x_0, R_*) \subset \Omega$,
where $R_* = \frac{3(1-a_0)}{3-5a_0}\eta$, then Eq. (3.16) admits a solution in $\overline{B(x_0, R_*)}$ and this solution is
unique in $B\left(x_0, \eta/a_0^2\right) \cap \Omega$.*

As $u(s) = 1$ in (3.16), $x_0(s) = u(s) = 1$ is a reasonable choice for $x_0(s)$. In this case,
the preceding result is valid for all λ such that

$$\frac{3\sqrt{2}|\lambda|^{3/2}}{(16 - 3|\lambda|)^{3/2}} \leq 0.3718\dots, \tag{3.18}$$

which is represented in Fig. 3.8, so that $|\lambda| \leq 1.98305\dots$

In particular, if we choose $\lambda = 1$, the previous inequality holds, $\beta = \frac{16}{13}$, $\eta = \frac{2}{13}$, and $K = \frac{3}{16}$. Therefore, $a_0 = K\beta\eta^{1/2} = 0.0905\ldots \leq \xi(1/2) = 0.3718\ldots$ and the hypotheses of Corollary 3.8 are satisfied. Then Eq. (3.16) with $\lambda = 1$ has a solution x^* in $\{v \in \mathcal{C}([0,1]) : \|v-1\| \leq 0.1647\ldots\} \subset \Omega$, which is unique in $\{v \in \mathcal{C}([0,1]) : \|v-1\| < 18.7778\ldots\} \cap \Omega$. Notice that this is an improvement over the existence domain obtained by Keller in [58], $\{v \in \mathcal{C}([0,1]) : \|v-1\| < 0.1842\ldots\}$.

3.3.3 Approximation of Solutions

Next, we transform the integral equation (3.16) with $\lambda = 1$ into a finite-dimensional problem. For this, we use the discretization process given in Sect. 1.2.2 and obtain that (3.16) is equivalent to the nonlinear system

$$x_i = 1 + \sum_{j=1}^{m} \alpha_{ij} x_j^{3/2}, \qquad i = 1, 2, \ldots, m,$$

where α_{ij}, for $i, j = 1, 2, \ldots, m$, are given in (1.11) with $a = 0$ and $b = 1$. This system can be then written as

$$\mathbb{F}(\mathbf{x}) \equiv \mathbf{x} - \mathbf{1} - A\,\hat{\mathbf{x}} = 0, \qquad \mathbb{F} : \mathbb{R}^m \longrightarrow \mathbb{R}^m, \qquad (3.19)$$

where $\mathbf{x} = (x_1, x_2, \ldots, x_m)^T$, $\mathbf{1} = (1, 1, \ldots, 1)^T$, $A = (\alpha_{ij})_{i,j=1}^{m}$, $\hat{\mathbf{x}} = \left(x_1^{3/2}, x_2^{3/2}, \ldots, x_m^{3/2}\right)^T$.

Since $x_0(s) = 1$ has been chosen as starting point for the theoretical study, a reasonable choice of initial approximation for Newton's method seems to be the vector $\mathbf{x}_0 = \mathbf{1}$. Then the conditions of Theorem 3.4 are satisfied. Indeed, formula (3.19) and the choice $m = 8$ yield

$$\mathbb{F}'(\mathbf{x}) = I - \frac{3}{2} A \operatorname{diag}\left\{x_1^{1/2}, x_2^{1/2}, \ldots, x_8^{1/2}\right\},$$

where I denotes the identity matrix. One can verify that

$$\|\mathbb{F}'(\mathbf{x}) - \mathbb{F}'(\mathbf{y})\| \leq K\|\mathbf{x} - \mathbf{y}\|^{1/2} \quad \text{with} \quad K = \frac{3}{2}\|A\| = 0.1853\ldots$$

Moreover,

$$\|[\mathbb{F}'(\mathbf{x}_0)]^{-1}\| \leq 1.2203\ldots = \beta, \qquad \|[\mathbb{F}'(\mathbf{x}_0)]^{-1}\mathbb{F}(\mathbf{x}_0)\| \leq 0.1468\ldots = \eta,$$

Table 3.1 Numerical solution
$\mathbf{x}^*$ of system (3.19)

i	x_i^*	i	x_i^*
1	1.01148126...	5	1.14780790...
2	1.05457678...	6	1.10966425...
3	1.10966425...	7	1.05457678...
4	1.14780790...	8	1.01148126...

Table 3.2 Absolute errors and
$\{\|\mathbb{F}(\mathbf{x}_n)\|\}$

n	$\|\mathbf{x}^* - \mathbf{x}_n\|$	$\|\mathbb{F}(\mathbf{x}_n)\|$
0	$1.4780\ldots \times 10^{-1}$	$1.2355\ldots \times 10^{-1}$
1	$9.1027\ldots \times 10^{-4}$	$7.6067\ldots \times 10^{-4}$
2	$3.2138\ldots \times 10^{-8}$	$2.6880\ldots \times 10^{-8}$

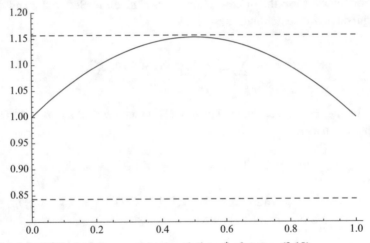

Fig. 3.9 Graph (solid line) of the approximate solution $\mathbf{x}^*$ of system (3.19)

and $a_0 = K\beta\eta^{1/2} = 0.0866\ldots < \xi(1/2) = 0.3718\ldots$, so that the conditions of Theorem 3.4 are satisfied. Therefore, Newton's method is convergent and, after three iterations, converges to the solution $\mathbf{x}^* = (x_1^*, x_2^*, \ldots, x_8^*)^T$ given in Table 3.1.

In Table 3.2 we show the errors $\|\mathbf{x}^* - \mathbf{x}_n\|$ and the sequence $\{\|\mathbb{F}(\mathbf{x}_n)\|\}$. This last sequence reveals that the vector shown in Table 3.1 is a good approximation of the solution of system (3.19) with $\mathrm{m} = 8$.

Finally, by interpolating the values of Table 3.1 and taking into account that the solution of (3.16) with $\lambda = 1$ satisfies $x(0) = 1 = x(1)$, we obtain the approximate solution drawn in Fig. 3.9. Observe also that this interpolated solution lies within the domain of existence of a solution, $\overline{B(\mathbf{1}, 0.1568\ldots)}$, provided by Theorem 3.4.

Operators with Hölder-Type Continuous First Derivative

4

The main aim of this chapter is to establish, using a modification of the condition (A2),

$$\|F'(x) - F'(y)\| \le L\|x - y\|, \quad x, y \in \Omega,$$

a new semilocal convergence result for Newton's method which modifies the domain of starting points obtained from the Newton-Kantorovich Theorem 2.1. In addition, in two particular cases, the new semilocal convergence result reduces to Theorem 2.1, stated for operators with Lipschitz continuous first Fréchet derivative, and Theorem 3.4, stated for operators with Hölder continuous first Fréchet derivative.

Next, we study the difficulties presented by the fulfilment of conditions (A1)–(A2)–(A3) of the Newton-Kantorovich Theorem 2.1 for some operators and propose a consistent alternative to condition (A2), prove the semilocal convergence of Newton's method under new convergence conditions and make some comments on the previous analysis in Sect. 4.2. Finally, we study nonlinear Hammerstein integral equations of the second kind; we determine where domains of existence and uniqueness of solutions, carry out a graphical analysis for a particular equation and approximate solutions by Newton's method when Theorem 2.1 cannot guarantee the semilocal convergence of the method.

We consider the Hammerstein integral equation given in (3.12),

$$x(s) = u(s) + \lambda \int_a^b \mathcal{K}(s, t)\, x(t)^q\, dt, \quad s \in [a, b], \quad \lambda \in \mathbb{R}, \quad q \in \mathbb{N}, \tag{4.1}$$

where $-\infty < a < b < +\infty$, $u(s)$ is a continuous function in $[a, b]$, the kernel $\mathcal{K}(s, t)$ is a known function in $[a, b] \times [a, b]$, and $x(s)$ is the unknown function.

© The Editor(s) (if applicable) and The Author(s), under exclusive licence to Springer Nature Switzerland AG 2020
J. A. Ezquerro Fernández, M. Á. Hernández Verón, *Mild Differentiability Conditions for Newton's Method in Banach Spaces*, Frontiers in Mathematics, https://doi.org/10.1007/978-3-030-48702-7_4

Observe that solving Eq. (4.1) is equivalent to solving the equation $\mathcal{F}(x) = 0$, where $\mathcal{F} : \Omega \subseteq \mathcal{C}([a, b]) \to \mathcal{C}([a, b])$ is given by

$$[\mathcal{F}(x)](s) = x(s) - u(s) - \lambda \int_a^b \mathcal{K}(s, t)\, x(t)^q dt, \quad s \in [a, b], \quad \lambda \in \mathbb{R}, \quad q \in \mathbb{N},$$

(4.2)

and Ω is to be determined. By an easy calculation

$$[\mathcal{F}'(x)y](s) = y(s) - q\lambda \int_a^b \mathcal{K}(s, t)x(t)^{q-1} y(t)\, dt,$$

(4.3)

$$[(\mathcal{F}'(x) - \mathcal{F}'(y))z](s) = -q\lambda \int_a^b \mathcal{K}(s, t) \left((x(t)^{q-1} - y(t)^{q-1} \right) z(t)\, dt.$$

(4.4)

As a consequence,

$$\|\mathcal{F}'(x) - \mathcal{F}'(y)\| \le q|\lambda|\ell \left(\|x\|^{q-2} + \|x\|^{q-3}\|y\| + \cdots + \|x\|\|y\|^{q-3} + \|y\|^{q-2} \right) \|x - y\|,$$

where $\ell = \max_{s \in [a,b]} \left| \int_a^b \mathcal{K}(s, t)\, dt \right|$.

Obviously, condition (A2) of the Newton-Kantorovich Theorem 2.1 is not satisfied in the present case if we cannot locate a solution of the equation $\mathcal{F}(x) = 0$ in some domain of the form $\Omega = B(v, \varrho)$ with $v \in \mathcal{C}([a, b])$ and $\varrho > 0$, where $\|x\|, \|y\| \le \|v\| + \varrho$. Thus, we can only calculate the constant L appearing in the theorem, $L = q(q - 1)|\lambda|\ell\, (\|v\| + \varrho)^{q-2}$, if we first locate a solution of $\mathcal{F}(x) = 0$.

In view of the condition (A3), $L\beta\eta \le \frac{1}{2}$, if the value of the parameter L is large, the starting point x_0 must be very close to a solution of our equation, so that condition (A3) is satisfied. For the equation $\mathcal{F}(x) = 0$ under study, it is clear that whether the condition (A3) is satisfied or not depends on the value of ϱ, which leads us to the need locate a starting point x_0 very close to the solution (and this is not easy).

On the other hand, in the worst case, if we cannot locate beforehand a solution of the equation, the aforementioned difficulty cannot be solved.

To avoid the two problems indicated above, we consider in this chapter the following condition:

$$\|F'(x) - F'(y)\| \le \omega(\|x\|, \|y\|)\, \|x - y\|^p, \quad x, y \in \Omega, \quad p \in (0, 1],$$

(4.5)

where $\omega : [0, +\infty) \times [0, +\infty) \to \mathbb{R}$ is a nondecreasing continuous function in both arguments and such that $\omega(0, 0) \ge 0$. Notice that we include the parameter p to consider operators with Hölder continuous Fréchet derivative: F' is Hölder continuous in Ω if $\omega(s, t) = K$.

Based on the condition (4.5), we prove the semilocal convergence of Newton's method by using a technique based on recurrence relations. It is required that the starting point x_0 is such that Newton's sequence $\{x_n\}$ is included in a ball $B(x_0, R) \subset \Omega$, $R > 0$. The conditions for the new semilocal convergence result do not depend directly on the domain Ω. This result pays attention to the existence of the value of R, which depends on the starting point x_0, but it avoids the difficulties implied by the Newton-Kantorovich Theorem 2.1. In addition, we provide relevant examples that clarify the different possibilities that may occur and observe the modification of the domain of starting points given by Theorem 2.1.

4.1 Convergence Analysis

We start by proving the semilocal convergence of Newton's method under condition (4.5). First, using the real parameters introduced by the conditions on the pair (F, x_0), we establish a system of recurrence relations where a sequence of positive real numbers is involved. This then allows us to guarantee the semilocal convergence of Newton's method in the Banach space X.

4.1.1 Recurrence Relations

We assume that

(B1) There exists the operator $\Gamma_0 = [F'(x_0)]^{-1} \in \mathcal{L}(Y, X)$, for some $x_0 \in \Omega$, with $\|\Gamma_0\| \leq \beta$ and $\|\Gamma_0 F(x_0)\| \leq \eta$.

(B2) There exists a function $\omega : [0, +\infty) \times [0, +\infty) \to \mathbb{R}$ such that $\|F'(x) - F'(y)\| \leq \omega(\|x\|, \|y\|) \|x - y\|^p$, $x, y \in \Omega$, $p \in (0, 1]$, nondecreasing and continuous in both arguments and $\omega(0, 0) \geq 0$.

In addition, we assume that the equation

$$\varphi(t) = \big((1 + p) - (2 + p)Q(t)\beta\eta^p\big) t - (1 + p)\big(1 - Q(t)\beta\eta^p\big) \eta = 0,$$

where $Q(t) = \omega(\|x_0\| + t, \|x_0\| + t)$, has at least one positive real root. We then denote the smallest such root by R.

Set $a_0 = Q(R)\beta\eta^p$ and define the scalar sequence $\{a_n\}$ by the rule

$$a_{n+1} = a_n f(a_n)^{1+p} g(a_n)^p, \quad n \geq 0, \tag{4.6}$$

where

$$f(t) = \frac{1}{1-t}, \qquad g(t) = \frac{t}{1+p}, \tag{4.7}$$

and we take $a_0 = 0$, so that we consider $a_0 > 0$.

Next, we prove the following recurrence relations for sequences (4.6) and $\{x_n\}$:

$$\|\Gamma_1\| = \|[F'(x_1)]^{-1}\| \le f(a_0)\|\Gamma_0\|, \tag{4.8}$$

$$\|x_2 - x_1\| \le f(a_0)g(a_0)\|x_1 - x_0\|, \tag{4.9}$$

$$Q(R)\|\Gamma_1\|\|x_2 - x_1\|^p \le a_1, \tag{4.10}$$

provided that

$$x_1 \in \Omega \qquad \text{and} \qquad a_0 < 1. \tag{4.11}$$

Indeed, if $x_1 \in \Omega$, then

$$\|I - \Gamma_0 F'(x_1)\| \le \|\Gamma_0\|\|F'(x_0) - F'(x_1)\| \le \beta\,\omega(\|x_0\|, \|x_1\|)\,\|x_1 - x_0\|^p \le Q(R)\beta\eta^p = a_0 < 1.$$

By the Banach lemma on invertible operators, there exists the operator $\Gamma_1 = [F'(x_0)]^{-1}$ and

$$\|\Gamma_1\| \le \frac{\|\Gamma_0\|}{1 - \|I - \Gamma_0 F'(x_1)\|} \le f(a_0)\|\Gamma_0\|.$$

Using a Taylor series expansion and the sequence $\{x_n\}$, we see that

$$\|F(x_1)\| = \left\|\int_0^1 (F'(x_0 + \tau(x_1 - x_0)) - F'(x_0))(x_1 - x_0)\,d\tau\right\|$$

$$\le \left(\int_0^1 \omega(\|x_0 + \tau(x_1 - x_0)\|, \|x_0\|)\,\tau^p dt\right)\|x_1 - x_0\|^{1+p}$$

$$\le \frac{Q(R)\eta^p}{1+p}\|x_1 - x_0\|,$$

and consequently

$$\|x_2 - x_1\| \le \|\Gamma_1\|\|F(x_1)\| \le f(a_0)g(a_0)\|x_1 - x_0\|,$$

$$Q(R)\|\Gamma_1\|\|x_2 - x_1\|^p \le a_0 f(a_0)^{1+p}g(a_0)^p = a_1.$$

Moreover, if $f(a_0)g(a_0) < 1$, then

$$\|x_2 - x_0\| \le (1 + f(a_0)g(a_0))\|x_1 - x_0\| < \frac{(1+p)(1-a_0)}{(1+p)-(2+p)a_0}\eta = R \qquad (4.12)$$

and $x_2 \in \Omega$ if $B(x_0, R) \subset \Omega$.

Later, we generalize the last recurrence relations to every term of the sequence $\{x_n\}$, and deduce that $\{x_n\}$ is a Cauchy sequence. For this, we first analyze the scalar sequence $\{a_n\}$.

4.1.2 Analysis of the Scalar Sequence

Let us verify that $\{x_n\}$ is a Cauchy sequence and (4.11) holds for all x_n and a_{n-1} with $n \ge 2$. First, we state a technical lemma whose proof is trivial.

Lemma 4.1 *Let f and g be the two real functions given in (4.7). Then*

(a) *f is increasing and $f(t) > 1$ in $(0, 1)$,*
(b) *g is increasing,*
(c) *for $\gamma \in (0, 1)$, we have $f(\gamma t) < f(t)$ if $t \in [0, 1)$ and $g(\gamma t) = \gamma g(t)$.*

Now we can prove some properties of the scalar sequence (4.6). For this, we consider the auxiliary function

$$\phi(t; p) = (1+p)^p(1-t)^{1+p} - t^p, \quad p \in (0, 1], \qquad (4.13)$$

which has only one zero, $\xi(p)$, in the interval $(0, 1/2]$, since $\phi(0; p) = (1+p)^p > 0$, $\phi(1/2; p) \le 0$ and $\phi'(t; p) < 0$ in $(0, 1/2]$. Notice that function (4.13) arises from setting the value $f(a_0)^{1+p}g(a_0)^p = \frac{a_0^p}{(1+p)^p(1-a_0)^{1+p}}$ equal to 1.

Lemma 4.2 *Let f and g be the two scalar functions defined in (4.7). If $a_0 \in (0, \xi(p))$, then*

(a) *$f(a_0)^{1+p}g(a_0)^p < 1$;*
(b) *the sequence $\{a_n\}$ is strictly decreasing;*
(c) *$a_n < 1$, for all $n \ge 0$.*

If $a_0 = \xi(p)$, then $a_n = a_0 < 1$ for all $n \ge 1$.

Proof We first consider the case $a_0 \in (0, \xi(p))$. Then, item (a) follows with strict inequality, since $\phi(a_0; p) > 0$. Item (b) is proved by induction on n. As $f(a_0)^{1+p} g(a_0)^p < 1$, we have $a_1 < a_0$. If we now suppose that $a_j < a_{j-1}$, for $j = 1, 2, \ldots, n$, then

$$a_{n+1} = a_n f(a_n)^{1+p} g(a_n)^p < a_n f(a_0)^{1+p} g(a_0)^p < a_n,$$

since f and g are increasing. Therefore, the sequence $\{a_n\}$ is strictly decreasing. To prove item (c), observe that $a_n < a_0 < 1$, for all $n \geq 0$, because the sequence $\{a_n\}$ is strictly decreasing and $a_0 \in (0, \xi(p))$.

Furthermore, if $a_0 = \xi(p)$, then $f(a_0)^{1+p} g(a_0)^p = 1$ and $a_n = a_0 = \xi(p) < 1$, for all $n \geq 0$. ∎

Lemma 4.3 *Let f and g be the two real functions given in (4.7). If $a_0 \in (0, \xi(p))$ and one denotes $\gamma = \frac{a_1}{a_0}$, then*

(a) $a_n < \gamma^{(1+p)^{n-1}} a_{n-1}$ *and* $a_n < \gamma^{\frac{(1+p)^n - 1}{p}} a_0$, *for all $n \geq 2$,*

(b) $f(a_n) g(a_n) < \gamma^{\frac{(1+p)^n - 1}{p}} f(a_0) g(a_0) = \dfrac{\gamma^{\frac{(1+p)^n}{p}}}{f(a_0)^{1/p}}$, *for all $n \geq 1$.*

If $a_0 = \xi(p)$, then $f(a_n) g(a_n) = f(a_0) g(a_0) = \frac{1}{f(a_0)^{1/p}}$, for all $n \geq 1$.

Proof We only prove the case $a_0 \in (0, \xi(p))$, since the case $a_0 = \xi(p)$ is deal with analogously. The proof of item (a) is done by induction. If $n = 2$, then by Lemma 4.1 (b)

$$a_2 = a_1 f(a_1)^{1+p} g(a_1)^p = \gamma a_0 f(\gamma a_0)^{1+p} g(\gamma a_0)^p < \gamma^{1+p} a_1 = \gamma^{2+p} a_0.$$

Now suppose that

$$a_{n-1} < \gamma^{(1+p)^{n-2}} a_{n-2} < \gamma^{\frac{(1+p)^{n-1} - 1}{p}} a_0.$$

Then, by the same reasoning,

$$a_n = a_{n-1} f(a_{n-1})^{1+p} g(a_{n-1})^p$$
$$< \gamma^{(1+p)^{n-2}} a_{n-2} f\left(\gamma^{(1+p)^{n-2}} a_{n-2}\right)^{1+p} g\left(\gamma^{(1+p)^{n-2}} a_{n-2}\right)^p$$
$$< \gamma^{(1+p)^{n-1}} a_{n-1}$$
$$< \cdots < \gamma^{\frac{(1+p)^n - 1}{p}} a_0.$$

To prove item (b), we observe that, for $n \geq 1$,

$$f(a_n)g(a_n) < f\left(\gamma^{\frac{(1+p)^n-1}{p}}a_0\right)g\left(\gamma^{\frac{(1+p)^n-1}{p}}a_0\right) < \gamma^{\frac{(1+p)^n-1}{p}}f(a_0)g(a_0) = \frac{\gamma^{\frac{(1+p)^n}{p}}}{f(a_0)^{1/p}}.$$

The proof is complete.   ∎

4.1.3   Semilocal Convergence Result

The equation $\varphi(t) = 0$ arises from imposing that all the points x_n are in the ball $B(x_0, R)$. In addition, we consider the following condition:

(B3) $a_0 = Q(R)\beta\eta^p \in (0, \xi(p)]$, where R is the smallest positive real root of the equation $\varphi(t) = 0$, $\xi(p)$ is the unique zero of function (4.13) in the interval $(0, 1/2]$, and $B(x_0, R) \subset \Omega$.

We are now ready to prove the semilocal convergence of Newton's method when it is applied to differentiable operators F such that F' satisfies a condition of type (4.5).

Theorem 4.4 *Let $F : \Omega \subseteq X \to Y$ be a continuously Fréchet differentiable operator defined on a nonempty open convex domain Ω of a Banach space X with values in a Banach space Y. Suppose that condition (B1)-(B2)-(B3) are satisfied. Then the Newton sequence $\{x_n\}$ starting at x_0 converges to a solution x^* of $F(x) = 0$ and $x_n, x^* \in B(x_0, R)$. Moreover, if equation*

$$\beta\,\omega(\|x_0\|, \|x_0\| + t)\left(t^{1+p} - R^{1+p}\right) = 1 + p$$

has at least one positive real root, and one denotes the smallest positive root by r, then x^* is unique in $B(x_0, r) \cap \Omega$. Furthermore, the sequence $\{x_n\}$ has R-order of convergence at least $1 + p$ if $a_0 \in (0, \xi(p))$, and at least one if $a_0 = \xi(p)$, and

$$\|x^* - x_n\| \leq \left(\gamma^{\frac{(1+p)^n-1}{p^2}}\right)\frac{\Delta^n}{1 - \gamma^{\frac{(1+p)^n}{p}}\Delta}\eta, \quad n \geq 0, \tag{4.14}$$

where $\gamma = \frac{a_1}{a_0}$ and $\Delta = (1 - a_0)^{1/p}$.

Proof We start with the case $a_0 \in (0, \xi(p))$. First, we prove that the following four recurrence relations are satisfied, for $n \geq 2$, for the sequence $\{x_n\}$:

(I_n) There exists the operator $\Gamma_{n-1} = [F'(x_{n-1})]^{-1}$ and $\|\Gamma_{n-1}\| \leq f(a_{n-2})\|\Gamma_{n-2}\|$;

(II_n) $\|x_n - x_{n-1}\| \leq f(a_{n-2})g(a_{n-2})\|x_{n-1} - x_{n-2}\|$;

(III_n) $Q(R)\|\Gamma_{n-1}\|\|x_n - x_{n-1}\|^p \leq a_{n-1}$;

(IV_n) $x_n \in \Omega$.

Since $\eta < R$ and R is the smallest positive zero of function φ, we have $x_1 \in \Omega$. By (4.8), (4.9), (4.10) and (4.12), we have that four relations hold for $n = 2$. If we now suppose that items (I_{n-1})–(II_{n-1})–(III_{n-1}) hold, then, by analogy to the case $n = 2$ we conclude that items (I_n)–(II_n)–(III_n) also hold.

Notice that $a_n < 1$ for all $n \geq 0$. Now, we prove (IV_n). Item (II_n) and item (b) of Lemma 4.3, imply exactly as in the proof of Theorem 3.4, that

$$\|x_n - x_0\| < \left(1 + \sum_{i=0}^{n-2}\left(\gamma^{\frac{(1+p)^{1+i}-1}{p^2}}\Delta^{1+i}\right)\right)\|x_1 - x_0\|,$$

where $\gamma = \frac{a_1}{a_0} < 1$ and $\Delta = \frac{f(a_0)g(a_0)}{\gamma^{1/p}} = \frac{1}{f(a_0)^{1/p}} = (1 - a_0)^{1/p} < 1$. By Bernoulli's inequality,

$$\gamma^{\frac{(1+p)^{1+i}-1}{p^2}} = \gamma^{1/p}\gamma^{\frac{1+p}{p^2}((1+p)^i-1)} \leq \gamma^{1/p}\gamma^{\frac{1+p}{p}i},$$

and therefore

$$\|x_n - x_0\| < \left(1 + \gamma^{1/p}\Delta\sum_{i=0}^{n-2}\gamma^{\frac{1+p}{p}i}\Delta^i\right)\|x_1 - x_0\| < \frac{\eta}{1 - \gamma^{1/p}\Delta} = R,$$

so that $x_n \in B(x_0, R)$. As $B(x_0, R) \subset \Omega$, then $x_n \in \Omega$, for all $n \geq 0$. Note that the conditions required in (4.11) are now satisfied for all x_n and a_{n-1} with $n \geq 2$.

Now let us prove that $\{x_n\}$ is a Cauchy sequence. Using item (II_n) and Lemma 4.3 (b), we have, in the same way as in the proof of Theorem 3.4, that

$$\|x_{n+m} - x_n\| < \sum_{i=0}^{m-1}\left(\gamma^{\frac{(1+p)^{n+i}-1}{p^2}}\Delta^{n+i}\right)\|x_1 - x_0\|.$$

Then, by Bernoulli's inequality,

$$\gamma^{\frac{(1+p)^{n+i}-1}{p^2}} = \gamma^{\frac{(1+p)^n-1}{p^2}}\gamma^{\frac{(1+p)^n}{p^2}((1+p)^i-1)} \leq \gamma^{\frac{(1+p)^n-1}{p^2}}\gamma^{\frac{(1+p)^n}{p}i}$$

and so

$$\|x_{n+m} - x_n\| < \frac{1 - \left(\gamma^{\frac{(1+p)^n}{p}} \Delta\right)^m}{1 - \gamma^{\frac{(1+p)^n}{p}} \Delta} \gamma^{\frac{(1+p)^n - 1}{p^2}} \Delta^n \eta. \tag{4.15}$$

Thus, $\{x_n\}$ is a Cauchy sequence, and hence there exists $x^* \in \overline{B(x_0, R)}$ such that $x^* = \lim_n x_n$.

Next, we prove that x^* is a solution of the equation $F(x) = 0$. Since $\|\Gamma_n F(x_n)\| \to 0$ as $n \to \infty$ and the sequence $\{\|F'(x_n)\|\}$ is bounded, because

$$\|F'(x_n)\| \leq \|F'(x_0)\| + \omega(\|x_0\|, \|x_n\|)\|x_n - x_0\|^p < \|F'(x_0)\| + \omega(\|x_0\|, \|x_0\| + R)R^p,$$

it follows that $\|F(x_n)\| \to 0$, as $n \to \infty$; indeed, $\|F(x_n)\| \leq \|F'(x_n)\|\|\Gamma_n F(x_n)\|$. Hence, by the continuity of F in $\overline{B(x_0, R)}$, we have $F(x^*) = 0$.

To prove the uniqueness of the solution x^*, suppose that y^* is another solution of $F(x) = 0$ in $B(x_0, r) \cap \Omega$. Then, from the approximation

$$0 = F(y^*) - F(x^*) = \int_{x^*}^{y^*} F'(x)dx = \int_0^1 F'(x^* + \tau(y^* - x^*)) \, d\tau (y^* - x^*),$$

it follows that $x^* = y^*$, provided that the operator $\int_0^1 F'(x^* + \tau(y^* - x^*)) \, d\tau$ is invertible. For this, we prove that, equivalently, there exists the operator J^{-1}, where $J = \Gamma_0 \int_0^1 F'(x^* + \tau(y^* - x^*)) \, d\tau$. Indeed, as

$$\|I - J\| \leq \|\Gamma_0\| \int_0^1 \|F'(x_0) - F'(x^* + \tau(y^* - x^*))\| \, d\tau$$

$$\leq \beta \int_0^1 \omega(\|x_0\|, \|x^* + \tau(y^* - x^*)\|) \|x_0 - (x^* + \tau(y^* - x^*))\|^p d\tau$$

$$\leq \beta \omega(\|x_0\|, \|x_0\| + r) \int_0^1 ((1 - \tau)\|x^* - x_0\| + \tau\|y^* - x_0\|)^p \, d\tau$$

$$< \beta \omega(\|x_0\|, \|x_0\| + r) \int_0^1 ((1 - \tau)R + \tau r)^p \, d\tau$$

$$= 1,$$

the operator J^{-1} exists thanks to the Banach lemma on invertible operators.

Finally, letting $m \to \infty$ in (4.15) we obtain (4.14) for all $n \geq 0$. Moreover, (4.14) implies that the R-order of convergence of the sequence $\{x_n\}$ is at least $1 + p$, since

$$\|x^* - x_n\| \leq \frac{\eta}{\gamma^{1/p^2}(1 - \gamma^{1/p}\Delta)} \left(\gamma^{1/p^2}\right)^{(1+p)^n}, \quad n \geq 0.$$

For the second case, $a_0 = \xi(p)$, we have $a_n = a_0 = \xi(p)$, for all $n \geq 0$. Then, taking into account that $\gamma = 1$ and $\Delta = f(a_0)g(a_0) < 1$ and following an analogous procedure, we obtain the same results except that the R-order of convergence is at least 1. ∎

4.2 A Little More

Next, we provide a convergence result for Newton's method when $p = 0$ in (4.5), a case that was not included in Sect. 4.1. After that, we comment on some particular cases of condition (B2). In addition, an analysis analogous to the one carried out for the scalar sequence (3.2) in Sect. 3.2.3, and different from that appearing in Sect. 4.1.2, can be given for sequence (4.6).

4.2.1 Case $p = 0$

If $p = 0$, the condition (B2) reduces

($\widetilde{B2}$) $\|F'(x) - F'(y)\| \leq \omega(\|x\|, \|y\|)$, $x, y \in \Omega$, where $\omega : [0, +\infty) \times [0, +\infty) \to \mathbb{R}$ is a nondecreasing and continuous function in both arguments and such that $\omega(0, 0) \geq 0$.

In this case, $a_0 = Q(R)\beta$, where $Q(t) = \omega(\|x_0\| + t, \|x_0\| + t)$ and R, if it exists, is the smallest positive real root of the equation $\varphi(t) = 0$ with $p = 0$. Proceeding in much the same way as in the case $p \in (0, 1]$, we see that

- if $a_0 < 1$ and $x_n \in \Omega$ $(n \geq 1)$, then $\|\Gamma_n\| \leq \dfrac{\|\Gamma_0\|}{1 - a_0}$, for all $n \geq 1$,
- $\|x_{n+1} - x_n\| \leq \left(\dfrac{a_0}{1 - a_0}\right)^n \|x_1 - x_0\|$, $n \geq 0$,

so that the semilocal convergence result for Newton's method is now the following.

Theorem 4.5 *Let $F : \Omega \subseteq X \to Y$ be a continuously Fréchet differentiable operator defined on a nonempty open convex domain Ω of a Banach space X with values in a Banach space Y. Suppose that conditions (B1)-($\widetilde{B2}$)-(B3) with $p = 0$ are satisfied. Suppose also that $a_0 = Q(R)\beta \in \left(0, \frac{1}{2}\right)$ and $B(x_0, R) \subset \Omega$. Then the Newton sequence $\{x_n\}$ starting at x_0 converges to a solution x^* of the equation $F(x) = 0$. Moreover,*

$x_n, x^* \in \overline{B(x_0, R)}$ *and* x^* *is unique in* $B(x_0, r) \cap \Omega$, *where* r, *if it exists, is the smallest positive real root of the equation* $\beta \, \omega(\|x_0\|, \|x_0\| + t) = 1$.

4.2.2   Particular Cases

Two well-known situations can be obtained as particular cases of condition (B2): operators with Lipschitz continuous first Fréchet derivative and operators with Hölder continuous first Fréchet derivative. If $\omega(s, t)$ is a constant L and $p = 1$ in the study carried out in Sect. 4.1, then F' is Lipschitz continuous in Ω and Theorem 4.4 reduces to the Newton-Kantorovich Theorem 2.1. If $\omega(s, t) = K$, then F' is Hölder continuous in Ω and Theorem 4.4 reduces to Theorem 3.4.

4.3   Application to a Hammerstein Integral Equation

Here we consider a nonlinear Hammerstein integral equation (4.1). First, we provide some results of existence and uniqueness of a solution. Second, we analyze a particular equation of (4.1) from a graphical point of view, and show that the domain of starting points given by the Newton-Kantorovich Theorem 2.1 is different from that in Theorem 4.4. Third, we locate solutions of the particular equation. And, finally, using Theorem 4.4, we guarantee the convergence of Newton's method and approximate two solutions.

4.3.1   Existence and Uniqueness of a Solution

We apply the study in Sect. 4.1 to obtain results on the existence and uniqueness of a solution of Eq. (4.1). We start by calculating the parameters β and η. If $x_0(s)$ is fixed, then (4.3) implies that

$$\|I - \mathcal{F}'(x_0)\| \leq q |\lambda| \|x_0^{q-1}\| \ell.$$

The Banach lemma on invertible operators ensures that, if $q |\lambda| \|x_0^{q-1}\| \ell < 1$, the operator $[\mathcal{F}'(x_0)]^{-1}$ exists and

$$\|[\mathcal{F}'(x_0)]^{-1}\| \leq \frac{1}{1 - q |\lambda| \|x_0^{q-1}\| \ell}.$$

From the definition of the operator $\mathcal{F}$ in (4.2) it follows that $\|\mathcal{F}(x_0)\| \leq \|x_0 - u\| + |\lambda| \|x_0^q\| \ell$ and

$$\|[\mathcal{F}'(x_0)]^{-1} \mathcal{F}(x_0)\| \leq \frac{\|x_0 - u\| + |\lambda| \|x_0^q\| \ell}{1 - q |\lambda| \|x_0^{q-1}\| \ell}.$$

On the other hand, thanks to (4.4), we have $\|\mathcal{F}'(x) - \mathcal{F}'(y)\| \le \omega(\|x\|, \|y\|)\|x - y\|$, where

$$\omega(s, t) = q|\lambda|\ell \left(s^{q-2} + s^{q-3}t + \cdots + st^{q-3} + t^{q-2} \right) \tag{4.16}$$

and $p = 1$. In addition, the equation $\varphi(t) = 0$ reduces to

$$(2 - 3Q(t)\beta\eta)\,t - 2\,(1 - Q(t)\beta\eta)\,\eta = 0 \tag{4.17}$$

and the function (4.13) to $\phi(t; 1) = 2t^2 - 5t + 2$, so that $\xi(1) = \frac{1}{2}$.

Once the parameters

$$\beta = \frac{1}{1 - q|\lambda|\|x_0^{q-1}\|\ell} \qquad \text{and} \qquad \eta = \frac{\|x_0 - u\| + |\lambda|\|x_0^q\|\ell}{1 - q|\lambda|\|x_0^{q-1}\|\ell}$$

are calculated and the function (4.16) is known, we can state the following result on the existence of solution of Eq. (4.1) from Theorem 4.4.

Theorem 4.6 *Let $\mathcal{F}$ be the operator defined in (4.2) and let $x_0 \in \Omega$ be a point such that there exists $[\mathcal{F}'(x_0)]^{-1}$. If $q|\lambda|\|x_0^{q-1}\|\ell < 1$, Eq. (4.17), where $Q(t) = \omega(\|x_0\|+t, \|x_0\|+t)$ and ω is defined in (4.16), has at least one positive real root, and the smallest positive real root, denoted by R, is such that $Q(R)\beta\eta \in \left(0, \frac{1}{2}\right]$ and $B(x_0, R) \subset \Omega$, then a solution of (4.1) exists at least in $\overline{B(x_0, R)}$. Moreover, if the equation*

$$\beta\,\omega(\|x_0\|, \|x_0\| + t)\left(t^2 - R^2 \right) = 2$$

has at least one positive real root and one denotes the smallest positive root by r, then x^* is unique in $B(x_0, r) \cap \Omega$.

Example 4.7 Consider the following particular equation of form (4.1)

$$x(s) = s + \frac{1}{2}\int_0^1 \mathcal{G}(s, t)\,x(t)^5\,dt, \tag{4.18}$$

where $\mathcal{G}(s, t)$ is the Green function in $[0, 1] \times [0, 1]$. Taking as initial approximation $x_0(s) = \theta u(s)$, with $\theta \in \mathbb{R}$, [3, 9], we see that the condition $q|\lambda|\|x_0^{q-1}\|\ell < 1$, that must be fulfilled for the corresponding operator $[\mathcal{F}'(x_0)]^{-1}$ to exist, reduces to $|\theta| < \sqrt[4]{\frac{16}{5}} = 1.3374\ldots$, since $q = 5$, $\lambda = \frac{1}{2}$, $\ell = \frac{1}{8}$, $u(s) = s$ and $\|x_0\| = |\theta|\|u\| = |\theta|$; see Fig. 4.1, where the last condition is drawn taking into account that the values of $\|x_0\| = |\theta|$ are represented on the horizontal axis and $q|\lambda|\|x_0^{q-1}\|\ell - 1 = \frac{5}{16}|\theta| - 1$ on the vertical axis.

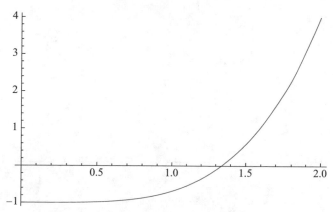

Fig. 4.1 $\|x_0\| = |\theta| < 1.3374\ldots$ for the existence of $[\mathcal{F}'(x_0)]^{-1}$

On the one hand, the Newton-Kantorovich for Theorem 2.1 to be applicable, we need first to locate a solution of integral equation (4.18) in order to obtain the value L required in the theorem. So, taking into account that a solution $x^*(s)$ of (4.18) in $\mathcal{C}([a, b])$ must satisfy

$$\|x^*\| - 1 - \frac{1}{16}\|x^*\|^5 \le 0,$$

it follows that $\|x^*\| \le \rho_1 = 1.1012\ldots$ or $\|x^*\| \ge \rho_2 = 1.5382\ldots$, where ρ_1 and ρ_2 are the two positive roots of the equation $\frac{t^5}{16} - t + 1 = 0$. Thus, using Theorem 2.1, we can approximate by Newton's method only one solution $x^*(s)$, which satisfies $\|x^*\| \in [0, \rho_1]$, since we can take $\Omega = B(0, \rho)$, with $\rho \in (\rho_1, \rho_2)$, where the corresponding operator $\mathcal{F}'(x)$ is Lipschitz continuous, and take $x_0 \in B(0, \rho)$ as starting point.

After that, taking into account what we have just mentioned, we can consider $\Omega = B(0, \rho)$ with $\rho \in (\rho_1, \rho_2)$, so that we then choose $\rho = \frac{3}{2}$. Then, as $\|x_0(s) - u(s)\| = |0 - 1|\|u(s)\| = |\theta - 1|$ in the upper bound of $\|[\mathcal{F}'(x_0)]^{-1}\mathcal{F}(x_0)\|$, we observe in Fig. 4.2, where Kantorovich's condition, $L\beta\eta \le \frac{1}{2}$, is drawn taking into account the values of $\|x_0\| = |\theta|$ are represented in the horizontal axis and $L\beta\eta - \frac{1}{2}$ in the vertical axis, that the starting point $x_0(s)$ is such that $\|x_0\| = |\theta| > 2.1710$ for condition $L\beta\eta \le \frac{1}{2}$ is satisfied. Therefore, it is clear that there is no starting point which satisfies the two conditions simultaneously, so that Theorem 2.1 of Newton-Kantorovich is not applicable in this situation.

On the other hand, if we consider Theorem 4.6, we see that, mainly, two conditions have to be fulfilled: first, the existence of at least one positive real root of Eq. (4.17); and second, the condition $Q(R)\beta\eta \in \left(0, \frac{1}{2}\right]$ on the smallest positive real root R of Eq. (4.17).

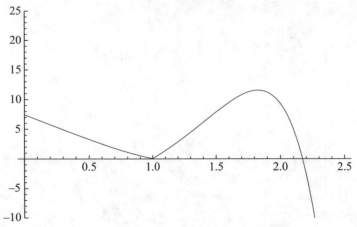

Fig. 4.2 $\|x_0\| = |\theta| \geq 2.1710\ldots$ for Kantorovich's condition $L\beta\eta \leq \frac{1}{2}$

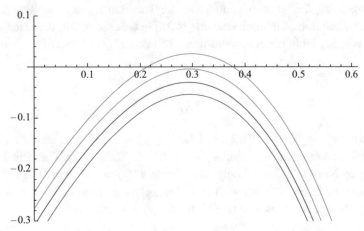

Fig. 4.3 $\varphi(t)$ with $\|x_0\| = |\theta| = 0.85, 0.875, 0.9, 0.925$ (blue, red, yellow and green, respectively)

Since the function (4.16) reduces to

$$\omega(s, t) = \frac{5}{16}\left(s^3 + s^2 t + s t^2 + t^3\right),$$

if we first focus our attention on the existence of at least one positive real root of Eq. (4.17), from Figs. 4.3 and 4.4, we suspect that $\|x_0\| = |\theta| > 0.9$ (more or less), so that Eq. (4.17) has positive real roots.

If we now focus our attention on the condition on R in Theorem 4.6, $Q(R)\beta\eta \leq \frac{1}{2}$, we see graphically in Figs. 4.5, 4.6, 4.7, and 4.8, where the curve is the function of

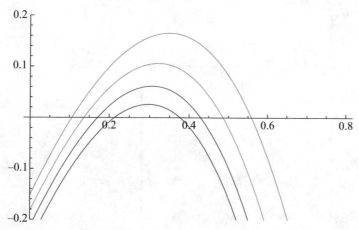

Fig. 4.4 $\varphi(t)$ with $\|x_0\| = |\theta| = 0.925, 0.95, 0.975, 1$ (blue, red, yellow and green, respectively)

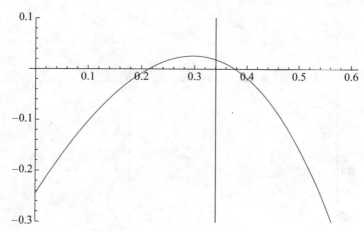

Fig. 4.5 $\varphi(t)$ with $\|x_0\| = |\theta| = 0.925$, $R = 0.2128\ldots$ and $Q(R)\beta\eta = 0.2948\ldots \leq \frac{1}{2}$

Eq. (4.17) and the vertical line is the function $Q(t)\beta\eta - \frac{1}{2}$, four different situations where the condition $Q(R)\beta\eta \leq \frac{1}{2}$ is satisfied.

Thus, we have exhibited four specific situations in which we can apply Theorem 4.6, but not the Newton-Kantorovich Theorem 2.1. In addition, the graphical analysis yields a modification of the domain of starting points for Newton's method given by Theorem 2.1.

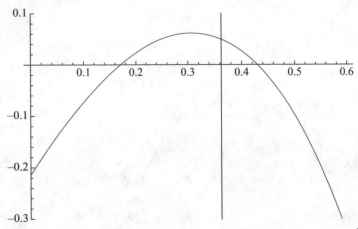

Fig. 4.6 $\varphi(t)$ with $\|x_0\| = |\theta| = 0.95$, $R = 0.1704\ldots$ and $Q(R)\beta\eta = 0.2921\ldots \leq \frac{1}{2}$

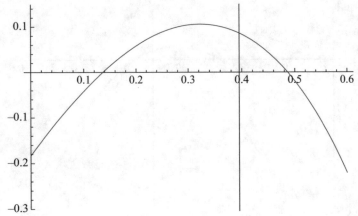

Fig. 4.7 $\varphi(t)$ with $\|x_0\| = |\theta| = 0.975$, $R = 0.1363\ldots$ and $Q(R)\beta\eta = 0.2668\ldots \leq \frac{1}{2}$

4.3.2   Localization of Solutions

Now, we consider the integral equation (4.18) and choose the starting point $x_0(s) = s$, as discussed in the previous section. Then, $\beta = \frac{16}{11}$, $\eta = \frac{1}{11}$ and the smallest real root of Eq. (4.17),

$$-(0.1517\ldots) + (1.5942\ldots)t - (1.3974\ldots)t^2 - (1.4575\ldots)t^3 - (0.4958\ldots)t^4 = 0,$$

is $R = 0.1062\ldots$ As a consequence, $Q(R) = 1.6921\ldots$, $Q(R)\beta\eta = 0.2237\ldots < \frac{1}{2}$ and the hypotheses of Theorem 4.6 are satisfied with $\Omega = \mathcal{C}([0, 1])$. Therefore, Eq. (4.18) has a solution $x^*(s)$ in the region $\{v \in \mathcal{C}([0, 1]) : \|v - s\| \leq 0.1062\ldots\}$ and is unique in $\{v \in \mathcal{C}([0, 1]) : \|v - s\| < 0.6685\ldots\}$.

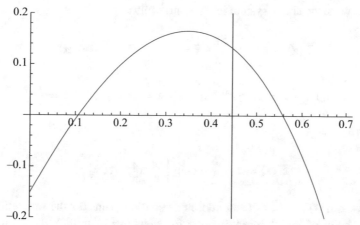

Fig. 4.8 $\varphi(t)$ with $\|x_0\| = |\theta| = 1$, $R = 0.1062\ldots$ and $Q(R)\beta\eta = 0.2237\ldots \leq \frac{1}{2}$

On the other hand, as the preceding graphic study revealed, if we consider $\Omega = B\left(0, \frac{3}{2}\right)$, then the condition $L\beta\eta \leq \frac{1}{2}$ of the Newton-Kantorovich Theorem 2.1 is not satisfied if $x_0(s) = s$, since $L\beta\eta = 0.5578\ldots > \frac{1}{2}$, so that we cannot guarantee the convergence of Newton's method to a solution of (4.18) based on Theorem 2.1. In addition, we cannot draw conclusions about the existence and uniqueness of a solution of Eq. (4.18) from Theorem 2.1.

Moreover, if we want to approximate a solution $x^{**}(s)$ such that $\|x^{**}(s)\| \geq \rho_2 = 1.5382\ldots$, we cannot apply Theorem 2.1 either because we cannot choose a domain where $x^{**}(s)$ lies; indeed, if the domain is chosen at random, it may not contain $x^{**}(s)$ or it may cut it, in which case $\mathcal{F}'(x)$ is not Lipschitz continuous.

In next section that we can consider the two previous situations from Theorem 4.6 if the hypotheses are assumed.

4.3.3   Approximation of Solutions

We apply Newton's method to approximate solutions with the features mentioned above. For this, we use the discretization process given in Sect. 1.2.2 with $m = 8$ and transform integral equation (4.18) into a finite-dimensional problem. Then, Eq. (4.18) is equivalent to the system of nonlinear equations

$$x_i = t_i + \frac{1}{2} \sum_{j=1}^{8} \alpha_{ij}\, x_j^5,$$

where α_{ij}, for $i, j = 1, 2, \ldots, 8$, are given in (1.11) with $a = 0$ and $b = 1$.

Then, we write the above system in the matrix form

$$\mathbb{F}(\mathbf{x}) \equiv \mathbf{x} - \mathbf{y} - \frac{1}{2} A \,\hat{\mathbf{x}} = 0, \qquad \mathbb{F} : \mathbb{R}^8 \longrightarrow \mathbb{R}^8, \qquad (4.19)$$

$$\mathbf{x} = (x_1, x_2, \ldots, x_8)^T, \quad \mathbf{y} = (t_1, t_2, \ldots, t_8)^T, \quad A = (\alpha_{jk})_{j,k=1}^8, \quad \hat{\mathbf{x}} = \left(x_1^5, x_2^5, \ldots, x_8^5\right)^T.$$

One can verify that

$$\mathbb{F}'(\mathbf{x}) = I - \frac{5}{2} A \,\mathrm{diag}\left\{x_1^4, x_2^4, \ldots, x_8^4\right\}.$$

Since we took $x_0(s) = s$ has been chosen as starting point for the theoretical study, a reasonable choice of initial approximation for Newton's method is the vector $\mathbf{x}_0 = \mathbf{y}$. After three iterations, we obtain the numerical approximation to the solution $\mathbf{x}^* = (x_1^*, x_2^*, \ldots, x_8^*)^T$ shown in Table 4.1. Observe that $\|\mathbf{x}^*\| = 0.9816\ldots \leq \rho_1 = 1.1012\ldots$

In Table 4.2, we show the errors $\|\mathbf{x}^* - \mathbf{x}_n\|$ and the sequence $\{\|\mathbb{F}(\mathbf{x}_n)\|\}$. This sequence demonstrates that the vector shown in Table 4.1 is a good approximation of the solution of system (4.19).

Thus, by interpolating the values of Table 4.1 and taking into account that a solution of (4.18) satisfies $x(0) = 0$ and $x(1) = 1$, we obtain the approximate solution drawn in Fig. 4.9. Observe that the interpolated approximation lies within the domain of existence of the solution obtained in Sect. 4.3.2.

We have seen previously that Eq. (4.18) may have a solution $x^{**}(s)$ such that $\|x^{**}(s)\| \geq \rho_2 = 1.5382\ldots$, but the convergence of Newton's method to this solution cannot be guaranteed based on the Newton-Kantorovich Theorem 2.1, since a domain where $x^{**}(s)$ lies and $\mathcal{F}'$ is Lipschitz continuous cannot be located. However, the convergence of Newton's method can be guaranteed by Theorem 4.6, as we can see in the following.

Table 4.1 Numerical solution $\mathbf{x}^*$ of the system (4.19)

i	x_i^*	i	x_i^*
1	0.02010275...	5	0.59887545...
2	0.10293500...	6	0.77066535...
3	0.24019309...	7	0.90392935...
4	0.41336490...	8	0.98160632...

Table 4.2 Absolute errors and $\{\|\mathbb{F}(\mathbf{x}_n)\|\}$

n	$\|\mathbf{x}^* - \mathbf{x}_n\|$	$\|\mathbb{F}(\mathbf{x}_n)\|$
0	$7.8991\ldots \times 10^{-3}$	$7.5595\ldots \times 10^{-3}$
1	$6.7668\ldots \times 10^{-6}$	$6.4814\ldots \times 10^{-6}$
2	$5.2591\ldots \times 10^{-12}$	$5.0502\ldots \times 10^{-12}$

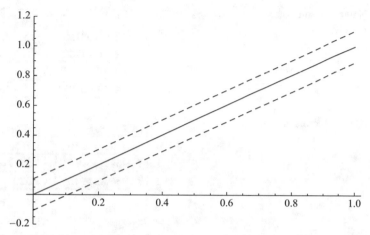

Fig. 4.9 Graph (solid line) of the approximate solution $\mathbf{x}^*$ of system (4.19)

If we choose the starting point $x_0(s) = 3s$, we find that $\|x_0(s)\| = 3 > \rho_2 = 1.5382\ldots$
In addition, a reasonable choice of initial approximation for Newton's method seems to be
the vector $\mathbf{x}_0 = 3\mathbf{y}$, but we cannot apply Theorem 4.6 either, since Eq. (4.17) cannot be
defined. However, it seems clear that the conditions of Theorem 4.6 can be satisfied if
the starting point is improved. So, starting at $\mathbf{x}_0$ and iterating with Newton's method, we
obtain the approximation

$$
\mathbf{z}_0 =
\begin{pmatrix}
0.081669\ldots \\
0.418183\ldots \\
0.975714\ldots \\
1.667559\ldots \\
2.199495\ldots \\
1.906819\ldots \\
1.406840\ldots \\
1.080048\ldots
\end{pmatrix},
$$

which is then used as a new starting point for Newton's method. For this new starting
point, Eq. (4.17) reduces to

$$
-(0.0108\ldots) + (0.9986\ldots)t - (1.3726\ldots)t^2 - (0.6250\ldots)t^3 - (0.0948\ldots)t^4 = 0,
$$

its smallest positive root is $R = 0.0110\ldots$ and the hypotheses of Theorem 4.6 are satisfied.
Observe that $\mathbf{z}_0$ is such that $\|\mathbf{z}_0\| = 2.1994\ldots > \rho_2 = 1.5382\ldots$ Choosing now the
starting point $\mathbf{z}_0$ and iterating again with Newton's method, we obtain the approximate
solution $\mathbf{x}^{**} = (x_1^{**}, x_2^{**}, \ldots, x_8^{**})^T$ given in Table 4.3, which is a solution that lies out

Table 4.3 Numerical solution of the system (4.19)

i	x_i^{**}	i	x_i^{**}
1	0.081556...	5	2.192901...
2	0.417603...	6	1.898637...
3	0.974358...	7	1.403096...
4	1.665051...	8	1.079315...

Table 4.4 Absolute errors and $\{\|\mathbb{F}(\mathbf{x}_n)\|\}$

n	$\|\mathbf{x}^{**} - \mathbf{x}_n\|$	$\|\mathbb{F}(\mathbf{x}_n)\|$
0	$8.1819\ldots \times 10^{-3}$	$1.7865\ldots \times 10^{-2}$
1	$7.9964\ldots \times 10^{-5}$	$1.5537\ldots \times 10^{-4}$
2	$7.2936\ldots \times 10^{-9}$	$1.5207\ldots \times 10^{-8}$

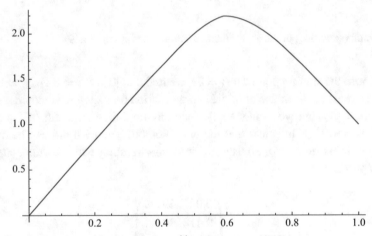

Fig. 4.10 Graph of the approximate solution $\mathbf{x}^{**}$ of the system (4.19)

of the grasp of the Newton-Kantorovich Theorem 2.1. Observe that $\|\mathbf{x}^{**}\| = 2.1929\ldots >$ $\rho_2 = 1.5382\ldots$

In Table 4.4 we show first few values of the errors $\|\mathbf{x}^{**} - \mathbf{x}_n\|$ and the sequence $\{\|\mathbb{F}(\mathbf{x}_n)\|\}$. From the last sequence, we observe that the vector shown in Table 4.3 is a good approximation of the solution of system (4.19).

Finally, by interpolating the values of Table 4.3 and taking into account that the solutions of (4.19) satisfy $x(0) = 0$ and $x(1) = 1$, we obtain the approximate solution drawn in Fig. 4.10.

Operators with ω-Lipschitz Continuous First Derivative

5

As we have seen in Chap. 3, when the conditions imposed on the operator F are relaxed with the hope than one can still prove the semilocal convergence of Newton's method, the following natural step is usually to consider operators with Hölder continuous first Fréchet derivative; i.e., assume that

$$\|F'(x) - F'(y)\| \le K\|x - y\|^p, \quad p \in [0, 1] \quad x, y \in \Omega.$$

Recall that F' is Lipschitz continuous in Ω if $p = 1$. In addition, we can also consider operators F with F' a combination of operators such that F' is Lipschitz or Hölder continuous in Ω, as, for example, the case for the Hammerstein integral equation

$$x(s) = u(s) + \int_a^b \mathcal{K}(s, t)\left(\lambda_1 x(t)^{1+p} s + \lambda_2 x(t)^q\right) dt, \quad p \in [0, 1], \quad q \in \mathbb{N}, \quad q \ge 2, \quad \lambda_1, \lambda_2 \in \mathbb{R}, \tag{5.1}$$

where $-\infty < a < b < +\infty$, $u(s)$ is a continuous function in $[a, b]$, the kernel $\mathcal{K}(s, t)$ is a known function in $[a, b] \times [a, b]$ and $x(s)$ is the unknown function.

We observe that solving the Eq. (5.1) is equivalent to solving the equation $\mathcal{F}(x) = 0$, where the operator $\mathcal{F} : \Omega \subseteq \mathcal{C}([a, b]) \to \mathcal{C}([a, b])$ is given by

$$[\mathcal{F}(x)](s) = x(s) - u(s) - \int_a^b \mathcal{K}(s, t)\left(\lambda_1 x(t)^{1+p} + \lambda_2 x(t)^q\right) dt, \quad s \in [a, b],$$

© The Editor(s) (if applicable) and The Author(s), under exclusive licence to Springer Nature Switzerland AG 2020
J. A. Ezquerro Fernández, M. Á. Hernández Verón, *Mild Differentiability Conditions for Newton's Method in Banach Spaces*, Frontiers in Mathematics,
https://doi.org/10.1007/978-3-030-48702-7_5

with $p \in [0, 1]$, $q \in \mathbb{N}$, $q \geq 2$ and $\lambda_1, \lambda_2 \in \mathbb{R}$. For this operator,

$$[\mathcal{F}'(x)y](s) = y(s) - \int_a^b \mathcal{K}(s, t) \left((1 + p)\lambda_1 x(t)^p + q\lambda_2 x(t)^{q-1} \right) y(t)\, dt$$

and

$$[(\mathcal{F}'(x) - \mathcal{F}'(y))z](s) =$$
$$- \int_a^b \mathcal{K}(s, t) \left((1 + p)\lambda_1 \left(x(t)^p - y(t)^p \right) + q\lambda_2 \left(x(t)^{q-1} - y(t)^{q-1} \right) \right) z(t)\, dt.$$

$$[(\mathcal{F}'(x) - \mathcal{F}'(y))z](s) =$$
$$- \int_a^b \mathcal{K}(s, t) \left((1 + p)\lambda_1 \left(x(t)^p - y(t)^p \right) + q\lambda_2 \left(x(t)^{q-1} - y(t)^{q-1} \right) \right) z(t)\, dt.$$

If $q = 2$, we have

$$\|\mathcal{F}'(x) - \mathcal{F}'(y)\| \leq \ell \left((1 + p)|\lambda_1| \|x - y\|^p + 2|\lambda_2| \|x - y\| \right),$$

where $\ell = \max\limits_{s \in [a,b]} \left| \int_a^b \mathcal{K}(s, t)\, dt \right|$. If $q \geq 3$, we have

$$\|\mathcal{F}'(x) - \mathcal{F}'(y)\| \leq \ell \big((1 + p)|\lambda_1| \|x - y\|^p$$
$$+ q|\lambda_2| \left(\|x\|^{q-2} + \|x\|^{q-3} \|y\| + \cdots + \|x\| \|y\|^{q-3} + \|y\|^{q-2} \right) \|x - y\| \big)$$
$$= \ell \left((1 + p)|\lambda_1| \|x - y\|^p + q|\lambda_2| \left(\sum_{i=0}^{q-2} \|x\|^{q-2-i} \|y\|^i \right) \|x - y\| \right).$$

This estimate shows that we have to bound $\|x\|$ and $\|y\|$, i.e., to fix the domain Ω. Since a solution $x^*(s)$ of the equation must be contained in Ω, the location of $x^*(s)$ is usually specified. From (5.1) it follows that

$$\|x^*\| \leq \|u\| + \ell\|\mathcal{H}(x^*)\| \leq U + \ell \left(|\lambda_1| \|x^*\|^{1+p} + |\lambda_2| \|x^*\|^q \right)$$

and

$$0 \leq \|u\| + \ell \left(|\lambda_1| \|x^*\|^{1+p} + |\lambda_2| \|x^*\|^q \right) - \|x^*\|. \tag{5.2}$$

This leads us to considering the scalar equation

$$\|u\| + \ell \left(|\lambda_1| \, t^{1+p} + |\lambda_2| \, t^q \right) - t = 0$$

and suppose that it has at least one positive solution; we denote its smallest positive solution by ρ_1. Obviously, in this case condition (5.2) is satisfied, provided that $\|x^*\| \leq \rho_1$. Hence, if the integral equation (5.1) has a solution $x^*(s) \in \overline{B(0, \rho_1)}$, then we can choose

$$\Omega = \{x \in \mathcal{C}([a, b]) : \; x(s) > 0, \; \|x\| < \rho, \; s \in [a, b]\},$$

for some $\rho > \rho_1$, and then Ω will be an open convex domain.

Note that, depending on what the exponents of $x(t)$ are in the integral equation (5.1), the condition $x(s) > 0$ in the definition of domain Ω can be omitted.

On the contrary, if we cannot guarantee the existence of a solution of the integral equation (5.1) in a ball $B(0, \rho)$, we cannot find a domain Ω that contains a solution $x^*(s)$ of (5.1) and where the condition (5.2) is satisfied.

To avoid the two problems described above, we consider in this chapter the generalized condition

$$\|F'(x) - F'(y)\| \leq \omega(\|x - y\|), \quad x, y \in \Omega, \tag{5.3}$$

where $\omega : [0, +\infty) \to \mathbb{R}$ is a continuous and nondecreasing function such that $\omega(0) \geq 0$. Obviously, if $\omega(z) = Lz$, then F' is Lipschitz continuous in Ω and, while if $\omega(z) = Kz^p$ with $p \in [0, 1]$, F' is Hölder continuous in Ω.

In what follows, we analyse the semilocal convergence of Newton's method under condition (5.3) and do it by using a technique based on recurrence relations that is similar to the one presented in Sect. 2.1 for Newton's method under condition (A2).

5.1   Convergence Analysis

In this section, we study the semilocal convergence of Newton's method to a unique solution of the equation $F(x) = 0$ under condition (5.3). For this, we construct a system of three recurrence relations that involved three sequences of positive numbers.

5.1.1   Recurrence Relations

We assume that:

(D1)   There exists the operator $\Gamma_0 = [F'(x_0)]^{-1} \in \mathcal{L}(Y, X)$, for some $x_0 \in \Omega$, with $\|\Gamma_0\| \leq \beta$ and $\|\Gamma_0 F(x_0)\| \leq \eta$.

(D2) $\|F'(x) - F'(y)\| \le \omega(\|x - y\|)$, $x, y \in \Omega$, where $\omega : [0, +\infty) \to \mathbb{R}$ is a continuous and nondecreasing function such that $\omega(0) \ge 0$.

(D3) There exists a continuous and nondecreasing function $h : [0, 1] \to \mathbb{R}$, such that $\omega(tz) \le h(t)\omega(z)$, for all $t \in [0, 1]$ and $z \in [0, +\infty)$.

Note that condition (D3) is not restrictive, since one can always take $h(t) = 1$, as ω is a nondecreasing function. We use it to sharpen the bounds that we obtain for particular expressions, as we will see later.

We now denote $a_0 = \eta$, $b_0 = \beta\omega(a_0)$ and define the following three scalar sequences:

$$t_n = I_h b_n f(b_n), \quad n \ge 0, \tag{5.4}$$

$$a_n = t_{n-1} a_{n-1}, \quad n \ge 1, \tag{5.5}$$

$$b_n = h(t_{n-1}) b_{n-1} f(b_{n-1}), \quad n \ge 1, \tag{5.6}$$

where $I_h = \displaystyle\int_0^1 h(t)\, dt$ and

$$f(t) = \frac{1}{1 - t}. \tag{5.7}$$

Obviously, we need to take $b_0 > 0$.

Let us show that the first terms of the sequences $\{x_n\}$, (5.5) and (5.6) satisfy the recurrence relations

$$\|\Gamma_1\| = \|[F'(x_1)]^{-1}\| \le f(b_0)\|\Gamma_0\|, \tag{5.8}$$

$$\|x_2 - x_1\| \le a_1, \tag{5.9}$$

$$\|\Gamma_1\|\omega(a_1) \le b_1. \tag{5.10}$$

For this, we assume that

$$x_1 \in \Omega \qquad \text{and} \qquad b_0 < 1.$$

Since the operator Γ_0 exists, the Banach lemma on invertible operators ensures that there exists the operator Γ_1 and

$$\|\Gamma_1\| \le \frac{\|\Gamma_0\|}{1 - \|I - \Gamma_0 F'(x_1)\|} \le f(b_0)\|\Gamma_0\|,$$

since

$$\|I - \Gamma_0 F'(x_1)\| \le \|\Gamma_0\| \|F'(x_0) - F'(x_1)\| \le \beta \omega(a_0) = b_0 < 1.$$

Using Taylor's formula and the definition of the sequence $\{x_n\}$, we obtain

$$\|F(x_1)\| = \left\| \int_0^1 (F'(x_0 + \tau(x_1 - x_0)) - F'(x_0))(x_1 - x_0) \, d\tau \right\|$$

$$\le \left(\int_0^1 \omega(\tau a_0) \, dt \right) \|x_1 - x_0\|$$

$$\le I_h \omega(a_0) \|x_1 - x_0\|.$$

Thus, since $t_0 < 1$,

$$\|\Gamma_1\| \omega(a_1) \le f(b_0) \|\Gamma_0\| \omega(t_0 a_0) \le f(b_0) \beta h(t_0) \omega(a_0) = h(t_0) b_0 f(b_0) = b_1,$$

$$\|x_2 - x_1\| \le \|\Gamma_1\| \|F(x_1)\| \le I_h b_0 f(b_0) \|x_1 - x_0\| \le t_0 a_0 = a_1,$$

$$\|x_2 - x_0\| \le \|x_2 - x_1\| + \|x_1 - x_0\| \le (t_0 + 1)a_0 < \frac{a_0}{1 - t_0},$$

and then $x_2 \in \Omega$, provided that $B(x_0, R) \subset \Omega$ with $R = \frac{a_0}{1-t_0}$.

Later, in Theorem 5.2 we generalize (5.8), (5.9) and (5.10) to every term of the sequence $\{x_n\}$ and prove that $\{x_n\}$ is a Cauchy sequence. To this end, we first investigate the scalar sequences $\{t_n\}$, $\{a_n\}$ and $\{b_n\}$ in the next subsection.

5.1.2 Analysis of the Scalar Sequences

Here we analyze the real sequences (5.4), (5.5) and (5.6) to guarantee the convergence of the sequence $\{x_n\}$. For this, it suffices to see that $x_{n+1} \in \Omega$ and $b_n < 1$, for all $n \ge 1$. Next, we generalize the recurrence relations introduced above, so that we can prove that $\{x_n\}$ is a Cauchy sequence. First, we give a technical lemma.

Lemma 5.1 *Let f be the scalar function in (5.7). If*

$$b_0 = \beta \omega(a_0) \le \frac{1}{1 + I_h} \qquad and \qquad b_0 < 1 - h(t_0), \qquad (5.11)$$

with $I_h = \displaystyle\int_0^1 h(t)\,dt$, then

(a) the sequences $\{t_n\}$, $\{a_n\}$ and $\{b_n\}$ are strictly decreasing;
(b) $t_n < 1$ and $b_n < 1$, for all $n \geq 0$.

If $b_0 = 1 - h(t_0) < \frac{1}{1+I_h}$, then $t_n = t_0 < 1$ and $b_n = b_0 < 1$, for all $n \geq 1$.

Proof First, suppose that (5.11) holds. Item (a) is proved by induction on n. As $b_0 < 1 - h(t_0)$, we have $b_1 < b_0$ and $t_1 < t_0$, because f is an increasing function. Moreover, $a_1 < a_0$ because $t_0 < 1$ and b_0 satisfies (5.11). Next, suppose that $b_i < b_{i-1}$, $t_i < t_{i-1}$ and $a_i < a_{i-1}$, for all $i = 1, 2, \ldots, n$. Then

$$b_{n+1} = h(t_n)b_n f(b_n) < h(t_0)b_n f(b_0) < b_n,$$

$$t_{n+1} = I_h b_n f(b_n) < I_h b_{n-1} f(b_{n-1}) = t_n,$$

$$a_{n+1} = t_n a_n < t_0 a_n < a_n,$$

since f and h are increasing functions in the interval $[0, 1)$. Consequently, the sequences $\{t_n\}$, $\{a_n\}$ and $\{b_n\}$ are strictly decreasing.

Second, for item (b), we have $t_n < t_0 < 1$ and $b_n < b_0 < 1$, for all $n \geq 0$, by item (a) and (5.11).

Finally, if $b_0 = 1 - h(t_0)$, then $h(t_0)f(b_0) = 1$ and, therefore, $b_n = b_0 = 1 - h(t_0) < 1$, for all $n \geq 0$. Moreover, if $b_0 < \frac{1}{1+I_h}$, then $t_n = t_0 < 1$, for all $n \geq 0$. ∎

5.1.3   Semilocal Convergence Result

We are now ready to prove a semilocal convergence result for Newton's method when it is applied to operators that satisfy the conditions (D1)-(D2)-(D3).

Theorem 5.2 Let $F : \Omega \subseteq X \to Y$ be a continuously Fréchet differentiable operator defined on a nonempty open convex domain Ω of a Banach space X with values in a Banach space Y. Suppose that (D1)-(D2)-(D3) are satisfied. Suppose also that (5.11) is satisfied and $B(x_0, R) \subset \Omega$, where $R = \frac{a_0}{1-t_0}$. Then the Newton sequence $\{x_n\}$ starting at x_0 converges to a solution x^* of $F(x) = 0$. Moreover, $x_n, x^* \in \overline{B(x_0, R)}$ and x^* is unique in $\Omega_0 = B(x_0, r) \cap \Omega$, where r is the biggest positive root of the equation

$$\frac{\beta}{r - R}\left(\Psi(r) - \Psi(R)\right) = 1, \qquad \Psi(r) = \int_0^r \omega(\tau)\,d\tau. \tag{5.12}$$

Proof First, we prove the following four recurrence relations for $n \geq 1$:

(I_n) There exists $\Gamma_n = [F'(x_n)]^{-1}$ and is such that $\|\Gamma_n\| \leq f(b_{n-1})\|\Gamma_{n-1}\|$;
(II_n) $\|x_{n+1} - x_n\| \leq a_n$;
(III_n) $\|\Gamma_n\|\omega(a_n) \leq b_n$;
(IV_n) $x_{n+1} \in \Omega$.

Notice that $x_1 \in \Omega$, since $\eta < R$. Then, from (5.8), (5.9), (5.10) and

$$\|x_2 - x_0\| \leq \|x_2 - x_1\| + \|x_1 - x_0\| \leq a_1 + a_0 = (t_0 + 1)a_0 < R,$$

it follows that (I_1)-(II_1)-(III_1)-(IV_1) hold. Now let us assume that (I_n)-(II_n)-(III_n)-(IV_n) hold for some $n = 1, 2, \ldots, i$, and show that (I_{i+1})-(II_{i+1})-(III_{i+1})-(IV_{i+1}) also hold. We use that $t_i < 1$ and $b_i < 1$, for all $i \geq 0$.

(I_{i+1}): Observe that

$$
\begin{aligned}
\|I - \Gamma_i F'(x_{i+1})\| &\leq \|\Gamma_i\|\omega(\|x_{i+1} - x_i\|) \\
&\leq f(b_{i-1})\|\Gamma_{i-1}\|\omega(a_i) \\
&= f(b_{i-1})\|\Gamma_{i-1}\|\omega(t_{i-1}a_{i-1}) \\
&\leq f(b_{i-1})\|\Gamma_{i-1}\|h(t_{i-1})\omega(a_{i-1}) \\
&\leq h(t_{i-1})b_{i-1}f(b_{i-1}) \\
&= b_i \\
&< 1,
\end{aligned}
$$

since $\{b_n\}$ is decreasing, $b_0 < \frac{1}{1+I_h}$ and $t_{i-1} < 1$. Then, by the Banach lemma on invertible operators, there exists Γ_{i+1} and

$$\|\Gamma_{i+1}\| \leq \frac{\|\Gamma_i\|}{1 - b_i} = f(b_i)\|\Gamma_i\|.$$

(II_{i+1}): Using a Taylor series expansion and the sequence $\{x_n\}$, we deduce, as for (5.9), that

$$
\begin{aligned}
\|F(x_{i+1})\| &= \left\| \int_0^1 (F'(x_i + \tau(x_{i+1} - x_i)) - F'(x_i))(x_{i+1} - x_i)\, d\tau \right\| \\
&\leq \left(\int_0^1 \omega(\tau\|x_{i+1} - x_i\|)\, d\tau \right) \|x_{i+1} - x_i\| \\
&\leq I_h\, \omega(a_i)\, a_i.
\end{aligned}
$$

Therefore,

$$\|x_{i+2} - x_{i+1}\| \le f(b_i)\|\Gamma_i\| I_h \omega(a_i)a_i \le I_h b_i f(b_i)a_i = t_i a_i = a_{i+1}.$$

(III_{i+1}): The inequality

$$\|\Gamma_{i+1}\| \omega(a_{i+1}) \le b_{i+1}$$

is immediate.

(IV_{i+1}): In addition, since the sequence $\{t_n\}$ is nonincreasing, we have

$$\|x_{i+2} - x_0\| \le \|x_{i+2} - x_{i+1}\| + \|x_{i+1} - x_0\|$$

$$\le a_{i+1} + \sum_{j=0}^{i} a_j$$

$$= a_0 \left(1 + \sum_{j=0}^{i} \left(\prod_{k=0}^{j} t_j \right) \right)$$

$$< a_0 \left(1 + \sum_{j=0}^{i} t_0^{j+1} \right)$$

$$= \frac{1 - t_0^{i+2}}{1 - t_0} a_0$$

$$< \frac{a_0}{1 - t_0}$$

$$= R,$$

so that $x_{i+2} \in B(x_0, R) \subset \Omega$. The induction is complete.

Second, taking into account item (II_n), we prove that $\{x_n\}$ is a Cauchy sequence. Indeed, for $m \ge 1$ and $n \ge 1$,

$$\|x_{n+m} - x_n\| \le \sum_{i=n}^{n+m-1} \|x_{i+1} - x_i\| \le \sum_{i=n}^{n+m-1} a_i \le a_0 t_0^n \frac{1 - t_0^m}{1 - t_0}.$$

Thus, $\{x_n\}$ is a Cauchy sequence and hence it is convergent.

Third, we see that $x^* = \lim_{n \to \infty} x_n$ is a solution of equation $F(x) = 0$. Indeed, the inequalities

$$\|F(x_n)\| \le \|F'(x_n)\| \|\Gamma_n F(x_n)\|,$$

$$\|F'(x_n)\| \le \|F'(x_0)\| + \omega(\|x_n - x_0\|) < \|F'(x_0)\| + \omega(R).$$

imply that $\|F(x_n)\| \to 0$ as $n \to \infty$, since $\|\Gamma_n F(x_n)\| \to 0$ by letting $n \to \infty$, and so $F(x^*) = 0$ by the continuity of F in $\overline{B(x_0, R)}$.

Finally, establish the uniqueness of x^*. Assume that y^* is another solution of $F(x) = 0$ in $\Omega_0 = B(x_0, r) \cap \Omega$. Then the approximation formula

$$0 = F(y^*) - F(x^*) = \int_{x^*}^{y^*} F'(x)dx = \int_0^1 F'(x^* + \tau(y^* - x^*))\, d\tau\, (y^* - x^*),$$

shows that $y^* = x^*$, provided that the operator $\int_0^1 F'(x^* + \tau(y^* - x^*))\, d\tau$ is invertible. The last is equivalent the operator $J = \Gamma_0 \int_0^1 F'(x^* + \tau(y^* - x^*))\, d\tau$ being invertible, which follows from the estimates

$$\|I - J\| \le \|\Gamma_0\| \int_0^1 \|F'(x^* + \tau(y^* - x^*)) - F'(x_0)\|\, d\tau$$

$$\le \beta \int_0^1 \omega(\|x_0 - x^* - \tau(y^* - x^*)\|)\, d\tau$$

$$\le \beta \int_0^1 \omega(\|(1 - \tau)(x_0 - x^*) - \tau(y^* - x_0)\|)\, d\tau$$

$$\le \beta \int_0^1 \omega((1 - \tau)\|x^* - x_0\| + \tau\|y^* - x_0\|)\, d\tau \qquad (5.13)$$

$$< \beta \int_0^1 \omega(R + \tau(r - R))\, d\tau$$

$$= \frac{\beta}{r - R} \int_R^r \omega(\tau)\, d\tau$$

$$= 1,$$

and the Banach lemma on invertible operators. ∎

Observe that in the preceding theorem the existence of the number r, which satisfies (5.12), is guaranteed if $\omega(R) < \dfrac{1}{2\beta \int_{1/2}^1 h(t)\, dt}$, since ω is a nondecreasing function. Moreover, r is unique. If this condition is not satisfied, since r does not exist.

From (5.13), it is easy to see that the uniqueness of the solution is guaranteed in $B(x_0, R)$ if $\omega(R) = 1/\beta$.

Remark 5.3 If $b_0 = 1 - h(t_0) < \frac{1}{1 + I_h}$, we can also prove, similarly to the previous theorem, that the Newton sequence converges in the Banach space X.

5.2   A Little More

In this section, we discuss the R-order of convergence of the method and comment on some particular cases that are included in the condition (5.3), in particular, the cases in which F' is Lipschitz continuous or Hölder continuous in Ω.

5.2.1   On the R-order of Convergence

We analyze the R-order of convergence of Newton's method when conditions (D1)-(D2)-(D3) are satisfied with $h(t) = t^d$, $t \in [0, 1]$ and $d \in [0, 1]$, so that $\omega(tz) \leq t^d \omega(z)$, for $z > 0$, $t \in [0, 1]$ and $d \in [0, 1]$. We will show that the R-order of convergence is at least $1 + d$.

Under the conditions (D1)-(D2)-(D3) with $h(t) = t^d$, $t \in [0, 1]$ and $d \in [0, 1]$, take $a_0 = \beta \omega(\eta)$ and observe that if $x_1 \in \Omega$ and $a_0 < 1$, then

$$\|I - \Gamma_0 F'(x_1)\| \leq \|\Gamma_0\| \|F'(x_0) - F'(x_1)\| \leq \beta \omega(\eta) = a_0 < 1.$$

By the Banach lemma on invertible operators, there exists the operator $\Gamma_1 = [F'(x_1)]^{-1}$ and $\|\Gamma_1\| \leq f(a_0) \|\Gamma_0\|$, where f is defined in (5.7). With the point x_1 is well defined.

Taylor's formula yields

$$\|F(x_1)\| = \left\| \int_0^1 (F'(x_0 + \tau(x_1 - x_0)) - F'(x_0))(x_1 - x_0) \, d\tau \right\|$$

$$\leq \left(\int_0^1 \omega(\tau \eta) \, d\tau \right) \|x_1 - x_0\|$$

$$\leq \frac{\omega(\eta)}{1 + d} \|x_1 - x_0\|,$$

so that

$$\|x_2 - x_1\| \leq \|\Gamma_1\| \|F(x_1)\| \leq \frac{a_0}{1 + d} f(a_0) \|x_1 - x_0\|,$$

$$\|x_2 - x_0\| \leq \|x_2 - x_1\| + \|x_1 - x_0\| \leq \left(1 + \frac{a_0}{1 + d} f(a_0) \right) \|x_1 - x_0\|,$$

$$\|\Gamma_1\| \omega(\|x_2 - x_1\|) \leq \beta f(a_0) \left(\frac{a_0}{1 + d} f(a_0) \right)^d \omega(\|x_1 - x_0\|) \leq \frac{a_0^{1+d}}{(1 + d)^d} f(a_0)^{1+d},$$

provided that $a_0 \leq \dfrac{1 + d}{2 + d}$, since f is increasing in $(0, 1)$.

Now set

$$\frac{a_0^{1+d}}{(1+d)^d} f(a_0)^{1+d} = a_1$$

and define the real sequence

$$a_n = \frac{a_{n-1}^{1+d}}{(1+d)^d} f(a_{n-1})^{1+d}, \quad n \geq 1.$$

Lemma 5.4 *If $a_0 = \beta\omega(\eta)$ satisfies*

$$a_0 \leq \frac{1+d}{2+d} \quad \text{and} \quad a_0^d < (1+d)^d(1-a_0)^{1+d}, \tag{5.14}$$

then

(a) *the sequence $\{a_n\}$ is decreasing,*

(b) $a_n \leq \dfrac{1+d}{2+d}$, *for all $n \geq 0$.*

The proof is obvious.

To obtain a priori error bounds for Newton's method when it converges to a solution x^* of the equation $F(x) = 0$, we first present a system of recurrence relations in the following lemma and some properties of the real sequence $\{a_n\}$ in Lemma 5.6.

Lemma 5.5 *If the conditions in (5.14) are satisfied, the following four statements hold for all $n \geq 1$:*

(I_n) *There exists the operator $\Gamma_n = [F'(x_n)]^{-1}$ and $\|\Gamma_n\| \leq f(a_{n-1})\|\Gamma_{n-1}\|$;*

(II_n) $\|x_{n+1} - x_n\| \leq \dfrac{a_{n-1}}{1+d} f(a_{n-1})\|x_n - x_{n-1}\|$;

(III_n) $\|\Gamma_n\|\omega(\|x_{n+1} - x_n\|) \leq a_n$;

(IV_n) $\|x_{n+1} - x_0\| \leq \dfrac{1 - \Delta^{n+1}}{1 - \Delta}\|x_1 - x_0\| < R\eta$, *where* $\Delta = \dfrac{a_0}{1+d} f(a_0)$ *and*

$R = \dfrac{1}{1 - \Delta}$

From a similar way to the above mentioned for $n = 1$ and using induction, the proof of the previous lemma follows.

Lemma 5.6 *Let f be the scalar function given by (5.7) and $\gamma = \frac{a_1}{a_0}$. If (5.14) is satisfied, then*

(a) $f(\gamma x) < f(x)$, *for $\gamma \in (0, 1)$ and $x \in (0, 1)$,*

(b) $a_n < \gamma^{(1+d)^{n-1}} a_{n-1}$ *and* $a_n < \gamma^{\frac{(1+d)^n - 1}{d}} a_0$, *for all $n \geq 2$.*

Proof Item (a) is obvious. The proof of item (b) is carried out by induction. Since $a_1 = \gamma a_0$ and $\gamma < 1$,

$$a_2 = \frac{a_1^{1+d}}{(1+d)^d} f(a_1)^{1+d} < \frac{(\gamma a_0)^{1+d}}{(1+d)^d} f(a_0)^{1+d} = \gamma^{1+d} a_1.$$

Now, suppose $a_{n-1} < \gamma^{(1+d)^{n-2}} a_{n-2} < \gamma^{\frac{(1+d)^{n-1}-1}{d}} a_0$. Then

$$a_n = \frac{a_{n-1}^{1+d}}{(1+d)^d} f(a_{n-1})^{1+d}$$

$$< \gamma^{(1+d)^{n-1}} \frac{a_{n-2}^{1+d}}{(1+d)^d} f(a_{n-2})^{1+d}$$

$$= \gamma^{(1+d)^{n-1}} a_{n-1}$$

$$< \gamma^{(1+d)^{n-1}} \gamma^{(1+d)^{n-2}} a_{n-2}$$

$$< \cdots < \gamma^{\frac{(1+d)^n-1}{d}} a_0,$$

as we needed. $\blacksquare$

Next, we prove a priori error bounds and calculate the R-order of convergence.

Theorem 5.7 *Let $F : \Omega \subseteq X \to Y$ be a continuously Fréchet differentiable operator defined on a nonempty open convex domain Ω of a Banach space X with values in a Banach space Y. Suppose that (D1)-(D2)-(D3) are satisfied with $h(t) = t^d$, $t \in [0, 1]$ and $d \in [0, 1]$. Suppose also that $a_0 = \beta \omega(\eta)$ satisfies (5.14) and $B(x_0, R) \subset \Omega$, where $R = \frac{1}{1-\Delta}$ and $\Delta = \frac{a_0}{1+d} f(a_0)$. Then, we have the following a priori error estimates:*

$$\|x^* - x_n\| \leq \left(\gamma^{\frac{(1+d)^n-1}{d^2}} \right) \frac{\Theta^n}{1 - \gamma^{\frac{(1+d)^n}{d}} \Theta} \eta, \quad n \geq 0, \tag{5.15}$$

where $\gamma = \frac{a_1}{a_0}$ and $\Theta = \frac{\Delta}{\gamma^{1/d}}$. Moreover, the Newton sequence has R-order of convergence at least $1 + d$.

Proof For $m \geq 1$, we have

$$\|x_{n+m} - x_n\| \leq \sum_{j=n}^{n+m-1} \|x_{j+1} - x_j\| \leq \|x_1 - x_0\| \sum_{j=n-1}^{n+m-2} \left(\prod_{i=0}^{j} \left(\frac{a_i}{1+d} f(a_i) \right) \right)$$

as a consequence of item (II_n) of Lemma 5.5. Since

$$\prod_{i=0}^{j} \frac{a_i}{1+d} f(a_i) = \Delta \, \Theta \, \gamma^{\frac{1+d}{d^2}((1+d)^j-1)}$$

and

$$\gamma^{\frac{(1+d)^{n+i}-1}{d^2}} = \gamma^{\frac{(1+d)^n-1}{d^2}} \gamma^{\frac{(1+d)^n}{d^2}((1+d)^i-1)} \le \gamma^{\frac{(1+d)^n-1}{d^2}} \gamma^{\frac{(1+d)^n}{d}i},$$

we see that

$$\|x_{n+m} - x_n\| \le \|x_1 - x_0\| \sum_{j=n-1}^{n+m-2} \left(\Delta \, \Theta^j \, \gamma^{\frac{1+d}{d^2}((1+d)^j-1)} \right)$$

$$= \|x_1 - x_0\| \sum_{i=0}^{m-1} \left(\Theta^{i+n} \, \gamma^{1/d} \, \gamma^{\frac{1+d}{d^2}((1+d)^{i+n-1}-1)} \right)$$

$$\le \|x_1 - x_0\| \, \Theta^n \, \gamma^{\frac{(1+d)^n-1}{d^2}} \sum_{i=0}^{m-1} \left(\Theta \gamma^{\frac{(1+d)^n}{d}} \right)^i.$$

Therefore,

$$\|x_{n+m} - x_n\| < \frac{1 - \left(\Theta \gamma^{\frac{(1+d)^n}{d}} \right)^m}{1 - \Theta \gamma^{\frac{(1+d)^n}{d}}} \, \Theta^n \, \eta \, \gamma^{\frac{(1+d)^n-1}{d^2}}.$$

Letting here $m \to \infty$, we obtain (5.15).

Now, from the estimates (5.15), it follows that

$$\|x^* - x_n\| \le \frac{\eta}{\gamma^{1/d^2}(1 - \Theta)} \left(\gamma^{1/d^2} \right)^{(1+d)^n}, \quad n \ge 0,$$

and consequently the R-order of convergence of the Newton sequence is at least $1 + d$. ∎

Remark 5.8 Observe that if F' is Hölder continuous in Ω, then $\omega(z) = K z^p$, $K \ge 0$, $p \in [0, 1]$ and $\omega(tz) \le t^p \omega(z)$. So, the R-order of convergence is at least $1 + p$, as already seen in Theorem 3.4.

5.2.2   Particular Cases

Extending the idea presented in the introduction to this chapter, we can find situations in which F' is as a combination of operators that are Lipschitz or Hölder continuous in Ω. In this case,

$$\|F'(x) - F'(y)\| \le \sum_{i=1}^{j} K_i \|x - y\|^{d_i}, \quad K_i \ge 0, \quad d_i \in [0, 1], \quad x, y \in \Omega,$$

so that we take $\omega(z) = \sum_{i=1}^{j} K_i z^{d_i}$, which satisfies

$$\omega(tz) = \sum_{i=1}^{j} K_i t^{d_i} z^{d_i} \le t^d \omega(z),$$

where $d = \max_{i=1,2,\ldots,j}\{d_i\}$, $d_i \in [0, 1]$, for $i = 1, 2, \ldots, j$, and $t \in [0, 1]$. In this case, the R-order of convergence is at least $1 + d$.

On the other hand, in (5.3) F' is Lipschitz continuous in Ω if $\omega(z) = Lz$ and Hölder continuous in Ω if $\omega(z) = Kz^p$ with $p \in [0, 1]$. If $\omega(z) = Lz$ in (5.3), the conditions required in Theorem 5.2 reduce to those in the Newton-Kantorovich Theorem 2.1, so that the semilocal convergence of Newton's method is guaranteed under the same hypotheses of Theorem 2.1. In addition, from Theorem 5.2, it follows that the solution x^* is unique in $\Omega_0 = B(x_0, r) \cap \Omega$, where $r = \frac{6b_0^2 - 18b_0 + 8}{3b_0(2 - 3b_0)}\eta$ and $b_0 = L\beta\eta$.

Finally, if $\omega(z) = Kz^p$ with $p \in (0, 1]$ in (5.3), then Theorem 5.2 reduces to Theorem 3.4, except for the uniqueness of the solution x^*, which now holds in $\Omega_0 = B(x_0, r) \cap \Omega$, where

$$r = \left(\frac{1 + p}{2K\beta(1 - 2^{-(1+p)})} \right)^{1/p} - R$$

and $R = \frac{(1+p)(1-b_0)}{(1+p)-(2+p)b_0}\eta$ with $b_0 = K\beta\eta^p$.

5.3   Application to a Hammerstein Integral Equation

Next, we provide some results of existence and uniqueness of a solution for Hammerstein integral equations of type (5.1). In particular, we consider (5.1) with $\lambda_1 = 1$, $\lambda_2 = \lambda$ and $q = 2$, i.e.,

$$x(s) = u(s) + \int_a^b \mathcal{K}(s, t) \left(x(t)^{1+p} + \lambda x(t)^2 \right) dt, \quad s \in [a, b], \quad p \in [0, 1], \quad \lambda \in \mathbb{R},$$

$$\tag{5.16}$$

where $-\infty < a < b < +\infty$, $u(s)$ is a continuous function in $[a, b]$, the kernel $\mathcal{K}(s, t)$ is a known function in $[a, b] \times [a, b]$ and $x(s)$ is the unknown function.

5.3.1 Existence and Uniqueness of a Solution

Observe that solving (5.16) is equivalent to solving the equation $\mathcal{F}(x) = 0$, where

$$\mathcal{F} : \Omega \subseteq \mathcal{C}([a, b]) \rightarrow \mathcal{C}([a, b]), \qquad \Omega = \{x \in \mathcal{C}([a, b]) : x(s) > 0, \ s \in [a, b]\},$$
(5.17)

$$[\mathcal{F}(x)](s) = x(s) - u(s) - \int_a^b \mathcal{K}(s, t) \left(x(t)^{1+p} + \lambda x(t)^2 \right) dt, \quad p \in [0, 1], \quad \lambda \in \mathbb{R}.$$
(5.18)

We apply the study of the preceding section to obtain results on the existence and uniqueness of solution of Eq. (5.16).

We start by calculating the parameters β and η that appear in the study. First, we have

$$[\mathcal{F}'(x)y](s) = y(s) - \int_a^b \mathcal{K}(s, t) \left((1 + p)x(t)^p + 2\lambda x(t) \right) y(t) \, dt.$$

Moreover, if $x_0(s)$ is fixed, then

$$\|I - \mathcal{F}'(x_0)\| \leq \left((1 + p)\|x_0^p\| + 2|\lambda|\|x_0\| \right) \ell.$$

By the Banach lemma on invertible operators, if

$$\left((1 + p)\|x_0^p\| + 2|\lambda|\|x_0\| \right) \ell < 1,$$

then

$$\|[\mathcal{F}'(x_0)]^{-1}\| \leq \frac{1}{1 - \left((1 + p)\|x_0^p\| + 2|\lambda|\|x_0\| \right) \ell}.$$

The definition of operator $\mathcal{F}$ implies that $\|\mathcal{F}(x_0)\| \leq \|x_0 - u\| + \left(\left\| x_0^{1+p} \right\| + |\lambda|\|x_0^2\| \right) \ell$ and

$$\|[\mathcal{F}'(x_0)]^{-1}\mathcal{F}(x_0)\| \leq \frac{\|x_0 - u\| + \left(\left\| x_0^{1+p} \right\| + |\lambda|\|x_0^2\| \right) \ell}{1 - \left((1 + p)\|x_0^p\| + 2|\lambda|\|x_0\| \right) \ell}.$$

On the other hand,

$$(\mathcal{F}'(x) - \mathcal{F}'(y))z(s) = -\int_a^b \mathcal{K}(s,t)((1+p)(x(t)^p - y(t)^p) + 2\lambda(x(t) - y(t)))z(t)\, dt$$

and $\|\mathcal{F}'(x) - \mathcal{F}'(y)\| \le \omega(\|x - y\|)$, where

$$\omega(z) = ((1+p)z^p + 2|\lambda|z)\, \ell. \tag{5.19}$$

Moreover, $\omega(tz) \le h(t)\omega(z)$, where $h(t) = t^p$, and $I_h = \displaystyle\int_0^1 h(t)\, dt = \dfrac{1}{1+p}$.

Once the parameters β and η are calculated and the function ω is known, we can derive from Theorem 5.2 the following result on the existence of a solution of the Eq. (5.16).

Theorem 5.9 *Let $\mathcal{F}$ be the operator defined in (5.17)–(5.18), and $x_0 \in \Omega$ be a point such that the operator $[\mathcal{F}'(x_0)]^{-1}$ exists. If $\big((1+p)\|x_0^p\| + 2|\lambda|\|x_0\|\big)\,\ell < 1$, (5.11) is satisfied, where ω is given by (5.19), $p \in [0,1]$, and $B(x_0, R) \subset \Omega$, where $R = \dfrac{(1+p)(1-b_0)}{(1+p)-(2+p)b_0}\eta$, then Eq. (5.16) has a solution at least in $\overline{B(x_0, R)}$. Moreover, this solution is unique in $B(x_0, r) \cap \Omega$, where r is the positive root of*

$$2\beta\omega(R + r)(2^{1+p} - 1) = (1+p)2^{1+p}.$$

Observe that the Newton sequence $\{x_n\}$ also converges if $b_0 = 1 - t_0^p < \frac{1+p}{2+p}$, see Remark 5.3.

Note also that the bound given for $\|\mathcal{F}(x_0)\|$ can be improved when the kernel $\mathcal{K}(s,t)$ and the function $u(s)$ are fixed.

5.3.2 Localization of Solutions

We consider a Hammerstein integral equation of the type (5.16) with $a = 0$, $b = 1$, $u(s) = 1$, $\lambda = \frac{1}{2}$, $p = \frac{1}{2}$ and $\mathcal{K}(s,t) = \mathcal{G}(s,t)$, where $\mathcal{G}(s,t)$ the Green function, given on $[0,1] \times [0,1]$, i.e.,

$$x(s) = 1 + \int_0^1 \mathcal{G}(s,t)\left(x(t)^{3/2} + \frac{1}{2}x(t)^2\right) dt, \quad s \in [0,1]. \tag{5.20}$$

If we choose $\Omega = \{x \in \mathcal{C}([0,1]) : x(s) > 0, \ s \in [0,1]\}$ and $x_0(s) = 1$, then

$$\beta = \frac{16}{11}, \quad \eta = \frac{3}{11}, \quad \omega(z) = \frac{1}{16}\left(3\sqrt{z} + 2z\right), \quad h(t) = \sqrt{t}, \quad R = 0.3240\ldots$$

As $\big((1+p)\|x_0^p\|+2|\lambda|\|x_0\|\big)\,\ell = \frac{5}{16} < 1$, $a_0 = \eta$, $b_0 = \beta\omega(\eta) = 0.1920\ldots \leq \frac{1+p}{2+p} = \frac{3}{5} = 0.6$, $b_0 = 0.1920\ldots < 1 - \Big(\frac{\beta\omega(\eta)}{(1+p)(1-\beta\omega(\eta))}\Big)^p = 0.6019\ldots$ and $B(x_0, R) = B(1, 0.3240\ldots) \subset \Omega$, the hypotheses of Theorem 5.9 are satisfied. As a consequence, (5.20) has a solution x^* in the region $\{v \in \Omega : \|v - 1\| \leq 0.3240\ldots\}$, which is unique in $\{v \in \Omega : \|v - 1\| < 3.2293\ldots\} \cap \Omega$.

5.3.3   Approximation of Solution

Next, we transform the integral equation (5.20) into a finite-dimensional problem by the discretization process given in Sect. 1.2.2 and arrive at the following equivalent nonlinear system:

$$x_i = 1 + \sum_{j=1}^{m} a_{ij}\left(x_j^{3/2} + \frac{1}{2}x_j^2\right), \quad j = 1, 2, \ldots, m,$$

where α_{ij}, for $i, j = 1, 2, \ldots, m$, are given in (1.11) with $a = 0$ and $b = 1$. This system is then written as

$$\mathbb{F}(\mathbf{x}) \equiv \mathbf{x} - \mathbf{1} - A\,\hat{\mathbf{x}} = 0, \qquad \mathbb{F} : \mathbb{R}^m \longrightarrow \mathbb{R}^m, \tag{5.21}$$

where $\mathbf{x} = (x_1, x_2, \ldots, x_m)^T$, $\mathbf{1} = (1, 1, \ldots, 1)^T$, $A = (\alpha_{ij})_{i,j=1}^m$, and

$$\hat{\mathbf{x}} = \left(x_1^{3/2} + \frac{1}{2}x_1^2, x_2^{3/2} + \frac{1}{2}x_2^2, \ldots, x_8^{3/2} + \frac{1}{2}x_m^2\right)^T.$$

Since we took $x_0(s) = 1$ as starting point for the theoretical study, a reasonable choice of initial approximation for Newton's method seems to be the vector $\mathbf{x}_0 = \mathbf{1}$. Let us verify that the conditions of Theorem 5.2 are satisfied in the discrete case. From (5.21) with $m = 8$, we obtain that the first derivative of the map $\mathbb{F}$ is

$$\mathbb{F}'(\mathbf{x}) = I - A\left(\frac{3}{2}\mathrm{diag}\left\{x_1^{1/2}, x_2^{1/2}, \ldots, x_8^{1/2}\right\} + \mathrm{diag}\{x_1, x_2, \ldots, x_8\}\right)$$

where I denotes the identity matrix. Consequently

$$\|\mathbb{F}'(\mathbf{x}) - \mathbb{F}'(\mathbf{y})\| \leq \omega(\|\mathbf{x} - \mathbf{y}\|) \quad \text{with} \quad \omega(z) = \frac{\|A\|}{2}\big(3\sqrt{z} + z\big) \quad \text{and} \quad h(t) = \sqrt{t}.$$

Moreover, as $\beta = 1.4198\ldots$ and $\eta = 0.2518\ldots$, we have $b_0 = \beta\omega(\eta) = 0.1541\ldots \leq \frac{1}{1+I_h} = 0.6$, $b_0 < 1 - h(t_0) = 0.6514\ldots$, $R = 0.2867\ldots$ and $B(1, 0.2867\ldots) \subset \Omega =$

Table 5.1 Numerical solution
$\mathbf{x}^*$ of the system (5.21)

i	x_i^*	i	x_i^*
1	1.01960381...	5	1.25935157...
2	1.09396463...	6	1.19091775...
3	1.19091775...	7	1.09396463...
4	1.25935157...	8	1.01960381...

Table 5.2 Absolute errors and
$\{\|\mathbb{F}(\mathbf{x}_n)\|\}$

n	$\|\mathbf{x}^* - \mathbf{x}_n\|$	$\|\mathbb{F}(\mathbf{x}_n)\|$
0	$2.5935\ldots \times 10^{-1}$	$1.8533\ldots \times 10^{-1}$
1	$7.4625\ldots \times 10^{-3}$	$5.2360\ldots \times 10^{-3}$
2	$6.1224\ldots \times 10^{-6}$	$4.2986\ldots \times 10^{-6}$
3	$4.1031\ldots \times 10^{-12}$	$2.8814\ldots \times 10^{-12}$

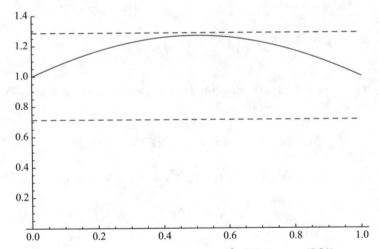

Fig. 5.1 Graph (solid line) of the approximate solution $\mathbf{x}^*$ of the system (5.21)

$\mathbb{R}^8$. Therefore, the hypotheses of Theorem 5.2 are satisfied. In addition, Newton's method converges to the solution $\mathbf{x}^* = (x_1^*, x_2^*, \ldots, x_8^*)^T$ shown in Table 5.1 after four iterations.

Table 5.2 shows the errors $\|\mathbf{x}^* - \mathbf{x}_n\|$ and the sequence $\{\|\mathbb{F}(\mathbf{x}_n)\|\}$. Examining this sequence, we see that the vector in Table 5.1 is a good approximation of a solution of the system (5.21) with $\mathfrak{m} = 8$.

Finally, by interpolating the values in Table 5.1 and taking into account that a solution of (5.20) satisfies $x(0) = 1 = x(1)$, we obtain the approximate solution drawn in Fig. 5.1. Observe also that this interpolated solution lies within the domain of existence of a solution, $B(\mathbf{1}, 0.2867\ldots)$, obtained from Theorem 5.2.

Recall that the best semilocal convergence result available for Newton's method

$$x_{n+1} = x_n - [F'(x_n)]^{-1} F(x_n), \quad n \geq 0, \quad \text{with } x_0 \text{ given,} \tag{6.1}$$

is the Newton–Kantorovich theorem, whose best-known variant is due to Ortega. In Sect. 2.1, we give a proof of this result based on recurrence relations under the following conditions:

(A1)   There exist the operator $\Gamma_0 = [F'(x_0)]^{-1} \in \mathcal{L}(Y, X)$, for some $x_0 \in \Omega$, with $\|\Gamma_0\| \leq \beta$ and $\|\Gamma_0 F(x_0)\| \leq \eta$.

(A2)   There exists a constant $L \geq 0$ such that $\|F'(x) - F'(y)\| \leq L\|x - y\|$ for $x, y \in \Omega$.

(A3) $a_0 = L\beta\eta \leq \frac{1}{2}$ and $B(x_0, R) \subset \Omega$, where $R = \frac{2(1-a_0)}{2-3a_0}\eta$.

Once the parameters β, η and L in (A1) and (A2) are known, the condition $a_0 = L\beta\eta \leq \frac{1}{2}$ in (A3) guarantees the semilocal convergence of Newton's method. If we consider the domain of parameters associated with the Newton–Kantorovich Theorem 2.1,

$$D = \left\{ (x, y) \in \mathbb{R}_+^2 : xy \leq \frac{1}{2} \right\},$$

we can draw D by choosing $x = L$ and $y = \beta\eta$ and coloring the region of the xy-plane specified by the condition $L\beta\eta \leq \frac{1}{2}$ of (A3), so that the convergence of Newton's method is guaranteed whenever one starts in D. See Fig. 6.2, where D is the black region. Observe that condition (A1), imposed to the initial approximation x_0, defines the parameters β and

© The Editor(s) (if applicable) and The Author(s), under exclusive licence to Springer Nature Switzerland AG 2020

J. A. Ezquerro Fernández, M. Á. Hernández Verón, *Mild Differentiability Conditions for Newton's Method in Banach Spaces*, Frontiers in Mathematics, https://doi.org/10.1007/978-3-030-48702-7_6

Fig. 6.1 Possible starting
points for Newton's method

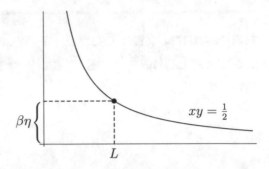

η, and condition (A2), imposed to the operator F, defines the parameter L fixed for all points of Ω.

Moreover (see [38]), to every point $z \in \Omega$ such that the operator $[F'(z)]^{-1}$ exists and satisfies $\|[F'(z)]^{-1}\| \leq \beta$ and $\|[F'(z)]^{-1}F(z)\| \leq \eta$ there is associated the pair $(L, \beta\eta)$ in the xy-plane, where $x = L$ and $y = \beta\eta$. Then, once the value of the constant L is fixed by condition (A2) for all points of Ω, we obtain a segment of possible points which represents the values of $\beta\eta$ which indicate that the corresponding starting points are good to obtain the convergence of Newton's method from them, as we can see in Fig. 6.1. Observe that every point $z \in \Omega$ such that the associated pair $(L, \beta\eta)$ belongs to D can be chosen as starting point for Newton's method, so that the method converges to a solution x^* of the equation $F(x) = 0$ when starting on it.

Recall that we can choose different values for the axes of the xy-plane depending on the study we do. In the previous case, we took $x = L$ and $y = \beta\eta$ because we were interested in the influence of the parameter L; namely, in the influence of the conditions on the first Fréchet derivative F' on the semilocal convergence of Newton's method (condition (A2) of Theorem 2.1).

In relation to the above, we expect that the larger the size of the domain of parameters the more possibilities for choosing good starting points for Newton's method will be available. As a consequence, we are interested in D is as large as possible, since this will allows us to find a larger number of good starting points for Newton's method.

The main aim of this chapter is to improve the domain of parameters D under conditions (A1)–(A2) or a generalization thereof. To modify the set D, one can use different strategies that change condition (A2), as shown in [12, 13, 17, 45]; however, these studies do not increase the size of D, they only modify it.

To increase the size of D and therefore the length of the specified segment, we propose to take into account that, as a consequence of condition (A2), once $x_0 \in \Omega$ is fixed, the condition

$$\|F'(x) - F'(x_0)\| \leq L_0\|x - x_0\|, \quad x \in \Omega, \tag{6.2}$$

is satisfied with $L_0 \leq L$. In what follows we denote $\mu = \frac{L_0}{L} \in (0, 1]$.

Then, by considering jointly the parameters L and L_0, we can relax the semilocal convergence conditions of Newton's method given in Theorem 2.1 by modifying condition (A3) and thus obtain a larger domain of parameters than D, so that the length of the segment shown in Fig. 6.1 is increased.

6.1   Operators with Lipschitz Continuous First Derivative

In this section, we follow [34] and improve the semilocal convergence of Newton's method for starting points in an increased domain D, under conditions on F', and using a technique based on recurrence relations which is similar to that presented in Sect. 2.1.

6.1.1   Convergence Analysis

We study the convergence of Newton's method to a unique solution x^* of the equation $F(x) = 0$ under some conditions on the pair (F, x_0). To this end, we construct, from certain real parameters, a system of four recurrence relations for a sequence of real positive numbers and use it to guarantee the convergence of Newton's method.

6.1.1.1   Recurrence Relations

From (A1), (A2) and (6.2), we denote $\theta_L = L\beta\eta$ and $\theta_L^0 = L_0\beta\eta = \mu\theta_L$, and then introduce the scalar sequence

$$a_1 = \theta_L f(\theta_L^0)^2 g(\theta_L^0), \qquad a_{n+1} = a_n f(a_n)^2 g(a_n), \qquad n \in \mathbb{N}, \tag{6.3}$$

where

$$f(t) = \frac{1}{1-t} \qquad \text{and} \qquad g(t) = \frac{t}{2}. \tag{6.4}$$

Note that an obvious problem results if $\theta_L = 0$, so we only take $\theta_L > 0$.

First, we prove the following four recurrence relations for the sequences (6.1) and (6.3):

$$\|\Gamma_1\| = \|[F'(x_1)]^{-1}\| \le f(\theta_L^0)\|\Gamma_0\|, \tag{6.5}$$

$$\|x_2 - x_1\| \le f(\theta_L^0)g(\theta_L^0)\|x_1 - x_0\|, \tag{6.6}$$

$$L\|\Gamma_1\|\|x_2 - x_1\| \le a_1, \tag{6.7}$$

$$\|x_2 - x_0\| \le (1 + f(\theta_L^0)g(\theta_L^0))\|x_1 - x_0\|, \tag{6.8}$$

provided that

$$x_1 \in \Omega \quad \text{and} \quad \theta_L^0 < 1. \tag{6.9}$$

If $x_1 \in \Omega$, then

$$\|I - \Gamma_0 F'(x_1)\| \le \|\Gamma_0\| \|F'(x_0) - F'(x_1)\| \le L_0 \beta \eta = \theta_L^0 < 1.$$

Now, the Banach lemma on invertible operators ensures that the operator Γ_1 exists and is such that

$$\|\Gamma_1\| \le \frac{\|\Gamma_0\|}{1 - \|I - \Gamma_0 F'(x_1)\|} \le f(\theta_L^0)\|\Gamma_0\|.$$

Then, using the Taylor series expansion and the sequence (6.1), we see that

$$\|F(x_1)\| = \left\| \int_0^1 (F'(x_0 + \tau(x_1 - x_0)) - F'(x_0))(x_1 - x_0)\, d\tau \right\| \le \frac{L_0}{2}\|x_1 - x_0\|^2.$$

Thus,

$$\|x_2 - x_1\| \le \|\Gamma_1\| \|F(x_1)\| \le f(\theta_L^0)g(\theta_L^0)\|x_1 - x_0\|$$

$$L\|\Gamma_1\| \|x_2 - x_1\| \le \theta_L f(\theta_L^0)^2 g(\theta_L^0) = a_1,$$

$$\|x_2 - x_0\| \le \left(1 + f(\theta_L^0)g(\theta_L^0)\right)\|x_1 - x_0\| < \left(1 + \frac{f(\theta_L^0)g(\theta_L^0)}{1 - f(a_1)g(a_1)}\right)\eta = R,$$

provided that $f(a_1)g(a_1) < 1$.

Second, we prove in the same way as above the following four recurrence relations for the sequences (6.1) and (6.3):

$$\|\Gamma_2\| = \|[F'(x_2)]^{-1}\| \le f(a_1)\|\Gamma_1\|, \tag{6.10}$$

$$\|x_3 - x_2\| \le f(a_1)g(a_1)\|x_2 - x_1\|, \tag{6.11}$$

$$L\|\Gamma_2\| \|x_3 - x_2\| \le a_2, \tag{6.12}$$

$$\|x_3 - x_0\| \le \left(1 + f(\theta_L^0)g(\theta_L^0)\left(1 + f(a_1)g(a_1)\right)\right)\|x_1 - x_0\|, \tag{6.13}$$

provided that

$$x_2 \in \Omega \quad \text{and} \quad a_1 < 1. \tag{6.14}$$

Third, we generalize theset recurrence relations to every element of the sequence (6.1), so as to guarantee that (6.1) is a Cauchy sequence. For this purpose, we first analyze the sequence $\{a_n\}$.

6.1.1.2 Analysis of the Scalar Sequence

Now, we analyze the scalar sequence defined in (6.3) in order to prove later the convergence of the sequence (6.1). For this, it suffices to verify that (6.1) is a Cauchy sequence and (6.9) and that (6.14) hold for all x_n and a_{n-1} with $n \geq 3$. We begin with a technical lemma whose proof is trivial.

Lemma 6.1 *Let f and g be the two real functions given in (6.4). Then*

(a) *f is increasing and $f(t) > 1$ in $(0, 1)$;*
(b) *g is increasing;*
(c) *for $\gamma \in (0, 1)$, we have $f(\gamma t) < f(t)$ if $t \in [0, 1)$ and $g(\gamma t) = \gamma g(t)$.*

Next, we give some properties of the scalar sequence $\{a_n\}$ defined in (6.3).

Lemma 6.2 *Let f and g be the two scalar functions defined in (6.4). If $\theta_L < \frac{(1-\mu\theta_L)^2}{\mu\theta_L}$, then*

(a) *$f(a_1)^2 g(a_1) < 1$;*
(b) *the sequence $\{a_n\}$ is strictly decreasing;*
(c) *$a_n < 1$, for all $n \geq 1$.*

If $\theta_L = \frac{(1-\mu\theta_L)^2}{\mu\theta_L}$, then $a_n = a_1 = \frac{1}{2}$ for all $n \geq 2$.

Proof We first consider the case $\theta_L < \frac{(1-\mu\theta_L)^2}{\mu\theta_L}$. Then, item (a) is obvious. Item (b) is proved by mathematical induction on n. As $f(a_1)^2 g(a_1) < 1$, we have $a_2 < a_1$. If we now suppose that $a_j < a_{j-1}$, for $j = 2, 3, \ldots, n$, then

$$a_{n+1} = a_n f(a_n)^2 g(a_n) < a_n f(a_1)^2 g(a_1) < a_n,$$

since f and g are increasing. Therefore, the sequence $\{a_n\}$ is strictly decreasing. To see item (c), we have $a_n < a_1 < 1$, for all $n \geq 1$, because the sequence $\{a_n\}$ is strictly decreasing and $\theta_L < \frac{(1-\mu\theta_L)^2}{\mu\theta_L}$.

Now, if $\theta_L = \frac{(1-\mu\theta_L)^2}{\mu\theta_L}$, then $f(a_1)^2 g(a_1) = 1$ and, consequently, $a_n = a_1 = \frac{1}{2} < 1$, for all $n \geq 1$. ∎

Lemma 6.3 *Let f and g be the two scalar functions defined in (6.4). Suppose $\theta_L < \frac{(1-\mu\theta_L)^2}{\mu\theta_L}$ and put $\gamma = \frac{a_2}{a_1}$. Then*

(a) $a_n < \gamma^{2^{n-2}} a_{n-1}$ *and* $a_n < \gamma^{2^{n-1}-1} a_1$, *for all $n \geq 3$;*

(b) $f(a_n)g(a_n) < \gamma^{2^{n-1}-1} f(a_1)g(a_1) = \dfrac{\gamma^{2^{n-1}}}{f(a_1)}$, *for all $n \geq 2$.*

If $\theta_L = \frac{(1-\mu\theta_L)^2}{\mu\theta_L}$, then $f(a_n)g(a_n) = f(a_1)g(a_1) = \frac{1}{f(a_1)}$, for all $n \geq 2$.

Proof We only prove the case $\theta_L < \frac{(1-\mu\theta_L)^2}{\mu\theta_L}$, since the case $\theta_L = \frac{(1-\mu\theta_L)^2}{\mu\theta_L}$ is dealt with in the same manner. The proof of item (a) is carried out by induction. If $n = 3$, then by the item (b) of Lemma 6.1, we have

$$a_3 = a_2 f(a_2)^2 g(a_2) = \gamma a_1 f(\gamma a_1)^2 g(\gamma a_1) < \gamma^2 a_1 f(a_1)^2 g(a_1) = \gamma^2 a_2.$$

Now suppose that

$$a_{n-1} < \gamma^{2^{n-3}} a_{n-2} < \gamma^{2^{n-2}-1} a_1.$$

Then, by the same reasoning,

$$a_n = a_{n-1} f(a_{n-1})^2 g(a_{n-1}) < \gamma^{2^{n-3}} a_{n-2} f\left(\gamma^{2^{n-3}} a_{n-2}\right)^2 < \gamma^{2^{n-2}} a_{n-1} < \cdots < \gamma^{2^{n-1}-1} a_1.$$

To prove item (b), we observe that, for $n \geq 2$,

$$f(a_n)g(a_n) < f\left(\gamma^{2^{n-1}-1} a_1\right) g\left(\gamma^{2^{n-1}-1} a_1\right) < \gamma^{2^{n-1}-1} f(a_1)g(a_1) = \dfrac{\gamma^{2^{n-1}}}{f(a_1)}.$$

The proof is complete. ∎

6.1.1.3 Semilocal Convergence Result

We are now ready to prove the semilocal convergence of Newton's method when F' satisfies condition (A2).

Theorem 6.4 *Let $F : \Omega \subseteq X \to Y$ be a continuously Fréchet differentiable operator defined on a nonempty open convex domain Ω of a Banach space X with values in a Banach space Y. Suppose that the conditions (A1)–(A2) are satisfied. If $\theta_L \leq \frac{(1-\mu\theta_L)^2}{\mu\theta_L}$*

and $B(x_0, R) \subset \Omega$, where $\mu = \frac{L_0}{L}$ and $R = \left(1 + \frac{f(\theta_L^0)g(\theta_L^0)}{1 - f(a_1)g(a_1)}\right)\eta$, then the Newton sequence (6.1) starting at x_0 converges to a solution x^* of the equation $F(x) = 0$ and $x_n, x^* \in \overline{B(x_0, R)}$, for all $n = 0, 1, 2, \ldots$ Moreover, we have the following a priori error bounds:

$$\|x^* - x_n\| \leq \gamma^{2^{n-1}-1} \frac{\Delta^{n-1}}{1 - \gamma^{2^{n-1}}\Delta}\|x_2 - x_1\|, \quad n \geq 2, \tag{6.15}$$

where $\gamma = \frac{a_2}{a_1}$ and $\Delta = (1 - a_1)^{-1}$.

Proof We begin with the case $\theta_L < \frac{(1-\mu\theta_L)^2}{\mu\theta_L}$. First, we prove that the sequence $\{x_n\}$ satisfies the following four conditions, for $n \geq 3$,

(I$_n$)   There exists $\Gamma_{n-1} = [F'(x_{n-1})]^{-1}$ and $\|\Gamma_{n-1}\| \leq f(a_{n-2})\|\Gamma_{n-2}\|$;
(II$_n$) $\|x_n - x_{n-1}\| \leq f(a_{n-2})g(a_{n-2})\|x_{n-1} - x_{n-2}\|$;
(III$_n$) $L\|\Gamma_{n-1}\|\|x_n - x_{n-1}\| \leq a_{n-1}$;
(IV$_n$) $x_n \in \Omega$.

First, $x_1 \in \Omega$, because $\eta < R$. Second, from (6.5)–(6.8), it follows that $x_2 \in \Omega$. Third, (6.10)–(6.13) imply items (I$_3$)–(II$_3$)–(III$_3$)–(IV$_3$). Now suppose that (I$_{n-1}$)–(II$_{n-1}$)–(III$_{n-1}$) hold. Then items (I$_n$)–(II$_n$)–(III$_n$) follow in the same way we prove (I$_3$)–(II$_3$)–(III$_3$). Notice that $a_n < 1$ for all $n \geq 1$. Now, let us prove (IV$_n$).

Observe that item (II$_n$) and Lemma 6.3 (b) imply that

$$\|x_n - x_0\| \leq \left(1 + \sum_{i=1}^{n-2}\left(\prod_{j=1}^{i} f(a_j)g(a_j)\right)\right)\|x_2 - x_1\| + \|x_1 - x_0\|$$

$$< \left(1 + \sum_{i=1}^{n-2}\left(\prod_{j=1}^{i} f(a_1)g(a_1)\gamma^{2^{j-1}-1}\right)\right)\|x_2 - x_1\| + \|x_1 - x_0\|$$

$$= \left(1 + \sum_{i=1}^{n-2}\left(\gamma^{2^i-1}\Delta^i\right)\right)\|x_2 - x_1\| + \|x_1 - x_0\|,$$

where $\gamma = \frac{a_2}{a_1} < 1$ and $\Delta = \frac{f(a_1)g(a_1)}{\gamma} = \frac{1}{f(a_1)} = 1 - a_1 < 1$. By the Bernoulli inequality, $\gamma^{2^i - 1} \leq \gamma^i$, and therefore

$$\|x_n - x_0\| < \left(1 + \sum_{i=1}^{n-2} (\gamma \Delta)^i\right) \|x_2 - x_1\| + \|x_1 - x_0\|$$

$$\leq \left(1 + f(\theta_L^0)g(\theta_L^0)\left(1 + \frac{\gamma \Delta}{1 - \gamma \Delta}\right)\right) \|x_1 - x_0\|$$

$$\leq R,$$

so that $x_n \in B(x_0, R)$. As $B(x_0, R) \subset \Omega$, we have $x_n \in \Omega$, for all $n \geq 0$. Note that the conditions required in (6.14) are now satisfied for all x_n and a_{n-1}, with $n \geq 3$.

Second, we prove that (6.1) is a Cauchy sequence. For this, we proceed as we did earlier. So, for $m \geq 2$ and $n \geq 2$, we have

$$\|x_{n+m} - x_n\| \leq \sum_{i=n-1}^{n+m-2} \left(\prod_{j=1}^{i} f(a_j)g(a_j)\right) \|x_2 - x_1\|$$

$$< \sum_{i=n-1}^{n+m-2} \left(\prod_{j=1}^{i} f(a_1)g(a_1)\gamma^{2^{j-1}-1}\right) \|x_2 - x_1\|$$

$$= \sum_{i=0}^{m-1} \left(\gamma^{2^{n+i-1}-1}\Delta^{n+i-1}\right) \|x_2 - x_1\|,$$

thanks to item (II_n) and Lemma 6.3 (b). By the Bernoulli inequality,

$$\gamma^{2^{n+i}-1} = \gamma^{2^{n-1}-1}\gamma^{2^{n-1}(2^i-1)} \leq \gamma^{2^{n-1}-1}\gamma^{2^{n-1}i},$$

so that

$$\|x_{n+m} - x_n\| < \left(\sum_{i=0}^{m-1} \left(\gamma^{2^{n-1}i}\Delta^i\right)\right) \gamma^{2^{n-1}-1}\Delta^{n-1}\|x_2 - x_1\|$$

$$< \frac{1 - \left(\gamma^{2^{n-1}}\Delta\right)^m}{1 - \gamma^{2^{n-1}}\Delta} \left(\gamma^{2^{n-1}-1}\right) \Delta^{n-1}\|x_2 - x_1\|. \tag{6.16}$$

Thus, (6.1) is a Cauchy sequence and then there exists $x^* \in \overline{B(x_0, R)}$ such that $x^* = \lim_n x_n$. Moreover, letting $m \to \infty$ in (6.16), we obtain (6.15).

Third, we prove that x^* is a solution of the equation $F(x) = 0$. As $\|\Gamma_n F(x_n)\| \to 0$ when $n \to \infty$, if we take into account that $\|F(x_n)\| \leq \|F'(x_n)\| \|\Gamma_n F(x_n)\|$ and $\{\|F'(x_n)\|\}$ is bounded, since

$$\|F'(x_n)\| \leq \|F'(x_0)\| + L_0\|x_n - x_0\| < \|F'(x_0)\| + L_0 R,$$

we conclued that $\|F(x_n)\| \to 0$ when $n \to \infty$. Hence, by the continuity of F in $\overline{B(x_0, R)}$, we have $F(x^*) = 0$. ∎

Notice that the condition $\theta_L \leq \frac{(1 - \mu\theta_L)^2}{\mu\theta_L}$ is equivalent to $0 \leq 1 - 2\mu\theta_L + \mu(\mu - 1)\theta_L^2$, which holds if $\theta_L \in \left[0, \frac{\sqrt{\mu} - \mu}{\mu(1 - \mu)}\right]$. In addition, from the variation of parameter μ, we have the following domains of parameters associated with Theorem 6.4:

$$D_{\text{Lip}}^\mu = \left\{(x, y) \in \mathbb{R}_+^2 : xy \leq \frac{\sqrt{\mu} - \mu}{\mu(1 - \mu)}\right\},$$

for each $\mu = \frac{L_0}{L} \in (0, 1]$.

6.1.2 Domain of Parameters

Recall that the accessibility of Newton's method can be analyzed by looking at domain of parameters associated with a semilocal convergence result and, in particular, with the Newton–Kantorovich theorem. If we now do the same with Theorem 6.4 and draw the associated domains of parameters, we can guess, from Fig. 6.2, that the smaller the quantity

Fig. 6.2 Domains of parameters of Newton's method associated with Theorem 2.1 (the black region) and Theorem 6.4

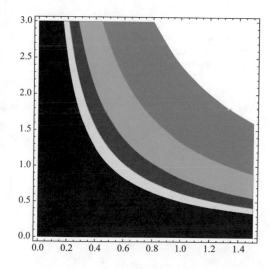

$\mu = \frac{L_0}{L} \in (0, 1]$, the bigger the domain of parameters: yellow for $\mu = 0.75$, red for $\mu = 0.5$, green for $\mu = 0.25$, and orange for $\mu = 0.1$. Moreover, as $\mu \to 1$, the domain of parameters associated with Theorem 6.4 tends to the one obtained by the Newton–Kantorovich theorem (the black region).

If we now compare the domains of parameters associated with Theorem 6.4 and the Newton–Kantorovich Theorem 2.1, D_{Lip}^{μ} and D, respectively, we see graphically that $D \subset D_{\text{Lip}}^{\mu}$, for all $\mu \in (0, 1]$. Thus, the domains of parameters associated with Theorem 6.4 improves the domain of parameters associated with the Newton–Kantorovich Theorem 2.1.

Moreover, we have the inclusions

$$D = D_{\text{Lip}}^1 = D_{\text{Lip}}^{\mu_j} \subset D_{\text{Lip}}^{\mu_{j-1}} \subset \cdots \subset D_{\text{Lip}}^{\mu_0} \qquad \text{for} \qquad \mu_0 < \cdots < \mu_{j-1} < \mu_j = 1.$$

On the other hand, we see the relationship between the domains of parameters associated with Theorems 2.1 and 6.4 from the variation of μ in Fig. 6.3, where we have chosen the values of μ on the x-axis and the values of $L\beta\eta$ in the y-axis. Figure 6.3 demonstrates that the domain of parameters associated with Theorem 6.4 (the gray region) is always larger than that associated with Theorem 2.1 (blue and gray regions) for all $\mu \in (0, 1]$.

Now, let us prove analytically what we see graphically. First, we prove that $D \subset D_{\text{Lip}}^{\mu}$, for each $\mu \in (0, 1]$. Indeed, if $(L, \beta\eta) \in D$, then $\theta_L = L\beta\eta \leq \frac{1}{2}$ and, as $\frac{\sqrt{\mu}-\mu}{\mu(1-\mu)} \geqslant \frac{1}{2}$ when $\mu \leq 1$, it follows that $(L, \beta\eta) \in D_{\text{Lip}}^{\mu}$ for all $\mu \in (0, 1]$.

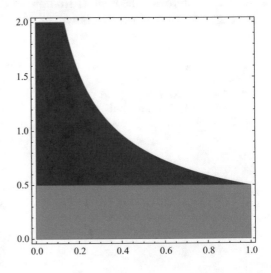

Fig. 6.3 Domains of parameters of Newton's method associated with Theorem 2.1 (the grey region) and Theorem 6.4 (blue and grey regions) from the variation of parameter μ

Fig. 6.4 Domain of parameters of Newton's method associated with the semilocal convergence result given by Argyros in [12]

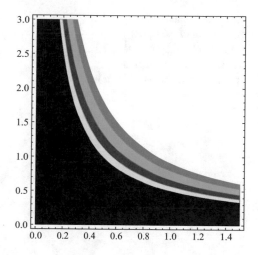

Second, we show that $D_{\text{Lip}}^{\mu_2} \subset D_{\text{Lip}}^{\mu_1}$ if $\mu_1 < \mu_2$ with $\mu_1, \mu_2 \in (0, 1]$. Indeed, from $\varphi(\mu) = \frac{(1-\mu\theta_L)^2}{\mu\theta_L}$, we have $\varphi'(\mu) = \frac{\theta_L}{\mu^2\theta_L^2}(\mu^2\theta_L^2 - 1) \leq \frac{\theta_L}{\mu^2\theta_L^2}(\mu^2 \frac{\sqrt{\mu}-\mu}{\mu(1-\mu)} - 1) < 0$, since $\mu^2 \frac{\sqrt{\mu}-\mu}{\mu(1-\mu)} < 1$, so that $\varphi(\mu_2) \leq \varphi(\mu_1)$ and, therefore, $D_{\text{Lip}}^{\mu_2} \subset D_{\text{Lip}}^{\mu_1}$.

On the other hand, in [12], I. K. Argyros also proves the semilocal convergence of Newton's method under (A1)–(A2) and using (6.2). For this, Argyros uses the majorant principle of Kantorovich [57]. As a consequence, the optimum case occurs when $L\beta\eta \leq \frac{1}{1+\mu}$. We can then see that the associated domain of parameters is

$$D_{\text{Ar1}}^{\mu} = \left\{ (x, y) \in \mathbb{R}_+^2 : xy \leq \frac{1}{1+\mu} \right\},$$

for each $\mu = L_0/L \in (0, 1]$, which is bigger than that associated with the Newton–Kantorovich Theorem 2.1, since $D \subset D_{\text{Ar1}}^{\mu}$ for $\mu < 1$, as we can see in Fig. 6.4, but smaller than that associated with Theorem 6.4, since $D_{\text{Ar1}}^{\mu} \subset D_{\text{Lip}}^{\mu}$ for $\mu < 1$. In addition, from Fig. 6.4, we can guess that the smaller the quantity $\mu = L_0/L \in (0, 1]$ is, the bigger the domain of parameters: yellow for $\mu = 0.75$, red for $\mu = 0.5$, green for $\mu = 0.25$ and orange for $\mu = 0.1$. Moreover, as $\mu \to 1$, the domain of parameters D_{Ar1}^{μ} tends to be that obtained from Theorem 2.1 (black region).

Argyros continued working in this way and obtained in [13] the domain of parameters given by

$$D_{\text{Ar2}}^{\mu} = \left\{ (x, y) \in \mathbb{R}_+^2 : xy \leq \frac{4}{1+4\mu+\sqrt{1+8\mu}} \right\},$$

Fig. 6.5 Domain of
parameters of Newton's
method: D, D_{Ar1}^{μ} and D_{Lip}^{μ}
(respectively: gray region,
yellow region and blue region)

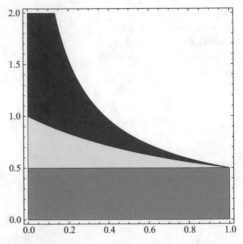

Fig. 6.6 Domain of
parameters of Newton's
method: D_{Ar1}^{μ}, D_{Ar2}^{μ}, D_{Lip}^{μ} and
D_{AH}^{μ} (respectively: yellow
region, green region, blue
region and cyan region)

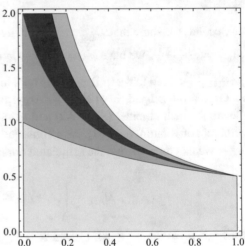

for each $\mu = L_0/L \in (0, 1]$, which was also obtained in [45] by other authors.
Additionally, Argyros and Hilout obtained in [14] the domain of parameters

$$D_{AH}^{\mu} = \left\{ (x, y) \in \mathbb{R}_+^2 : xy \leq \frac{4}{4\mu + \sqrt{\mu + 8\mu^2} + \sqrt{\mu}} \right\},$$

for each $\mu = L_0/L \in (0, 1]$. In order to compare these new domains of parameters with
that associated with Theorem 6.4, we rewrite them in terms of the variation of μ, see
Figs. 6.5 and 6.6, where the values of μ are represented along the x-axis and the values
of $L\beta\eta$ along the y-axis. Figure 6.5 confirms that $D \subset D_{Ar1}^{\mu} \subset D_{Lip}^{\mu}$ and Fig. 6.6 does
$D_{Ar1}^{\mu} \subset D_{Ar2}^{\mu} \subset D_{Lip}^{\mu} \subset D_{AH}^{\mu}$.

However, as we will see at the end of Sect. 6.1.3, in the presented application, the use of recurrence relations allows us to obtain better a priori error estimates than those that would be obtained by Argyros and Hilout's approach with majorizing sequences. They consider the majorant sequence $\{t_n\}$,

$$t_0 = 0, \quad t_1 = \eta, \quad t_2 = \eta + \frac{L_0 \beta \eta^2}{2(1 - L_0 \beta \eta)}, \quad t_{n+2} = t_{n+1} + \frac{L\beta(t_{n+1} - t_n)^2}{2(1 - L_0 \beta t_{n+1})}, \quad n \in \mathbb{N},$$

to prove the semilocal convergence of Newton's method in [14] and use it to calculate the a priori error estimates given in the application of Sect. 6.1.3.

6.1.3 Application to a Conservative Problem

It is well known that, in the evolution of any real dynamical system, energy is dissipated, usually through some form of friction. However, in certain situations this dissipation is so slow that it can be neglected over relatively short periods of time. In such cases, we assume the law of conservation of energy, namely, that the sum of the kinetic energy and the potential energy is constant. A system of this kind is said to be conservative.

If Φ and Υ are arbitrary functions such that $\Phi(0) = 0$ and $\Upsilon(0) = 0$, then the general equation

$$\lambda \frac{d^2 x(t)}{dt^2} + \Upsilon \left(\frac{dx(t)}{dt} \right) + \Phi(x(t)) = 0 \tag{6.17}$$

can be interpreted as the equation of motion of a mass λ under the action of a restoring force $-\Phi(x)$ and a damping force $-\Upsilon(dx/dt)$. In general, these forces are nonlinear, and (6.17) can be regarded as the basic equation of nonlinear mechanics. Now, let us consider the special case of a nonlinear conservative system described by the equation

$$\lambda \frac{d^2 x(t)}{dt^2} + \Phi(x(t)) = 0,$$

in which the damping force is zero and consequently there is no dissipation of energy. Extensive discussions of (6.17), with applications to a variety of physical problems, can be found in the classical references [4] and [74].

Now, let us consider the special case of a nonlinear conservative system described by the equation

$$\frac{d^2 x(t)}{dt^2} + \Psi(x(t)) = 0 \tag{6.18}$$

with the boundary conditions

$$x(0) = c_1, \quad x(1) = c_2. \tag{6.19}$$

We will use a discretization process to transform problem (6.18)–(6.19) into a finite-dimensional problem and look for an approximation of a solution when a particular function $\Psi(x(t))$ is considered. Specifically, we transform problem (6.18)–(6.19) into a system of nonlinear equations by approximating the second derivative by a standard numerical formula.

6.1.3.1 Setting up a Finite Difference Scheme

Firstly, we introduce the points $t_j = j\kappa$, $j = 0, 1, \ldots, \mathfrak{m}+1$, where $\kappa = \frac{1}{\mathfrak{m}+1}$ and $\mathfrak{m}$ is an appropriate integer. A scheme is then designed for the determination of numbers x_j that approximate the values $x(t_j)$ of the true solution at the points t_j. A standard approximation for the second derivative at these points is

$$x_j'' \approx \frac{x_{j-1} - 2x_j + x_{j+1}}{\kappa^2}, \quad j = 1, 2, \ldots, \mathfrak{m}.$$

A natural way to obtain such a scheme is to demand that the variables x_j satisfy at the interior mesh points t_j the difference equations

$$x_{j-1} - 2x_j + x_{j+1} + \kappa^2 \Psi(x_j) = 0, \quad j = 1, 2, \ldots, \mathfrak{m}. \tag{6.20}$$

Since x_0 and $x_{\mathfrak{m}+1}$ are determined by the boundary conditions, the unknowns are $x_1, x_2, \ldots, x_{\mathfrak{m}}$.

The discussion is further simplified by the use of matrix and vector notation. Introducing the vectors

$$\mathbf{x} = (x_1, x_2, \ldots, x_{\mathfrak{m}})^T, \quad v_{\mathbf{x}} = (\Psi(x_1), \Psi(x_2), \ldots, \Psi(x_{\mathfrak{m}}))^T$$

and the matrix

$$A = \begin{pmatrix} -2 & 1 & 0 & \cdots & 0 \\ 1 & -2 & 1 & \cdots & 0 \\ 0 & 1 & -2 & \cdots & 0 \\ \vdots & \vdots & \vdots & \ddots & \vdots \\ 0 & 0 & 0 & \cdots & -2 \end{pmatrix}_{\mathfrak{m} \times \mathfrak{m}},$$

the system of Eqs. (6.20) can be written compactly in the form

$$\mathbb{F}(\mathbf{x}) \equiv A\mathbf{x} + \kappa^2 v_{\mathbf{x}} = 0, \quad \mathbb{F} : \mathbb{R}^{\mathfrak{m}} \longrightarrow \mathbb{R}^{\mathfrak{m}}. \tag{6.21}$$

From now on, we focus on solving the system (6.21) with $m = 8$.

Example 6.5 The steady temperature distribution is known in a homogeneous rod of length 1 in which, as a consequence of a chemical reaction or some other heat-producing process, heat is generated at a rate $\Psi(x(t))$ per unit time per unit length, $\Psi(x(t))$ being a given function of the excess temperature x of the rod over the temperature of the surroundings. We assume that the ends of the rod, $t = 0$ and $t = 1$, are kept at given temperatures, discretize the boundary value problem (6.18)–(6.19) along the axis of the rod, and approximate a solution of a nonlinear system (6.21) that arises from the discretization process.

For an example, we consider the law $\Psi(x(t)) = 1 + x(t)^3$ for heat generation. Thus, we consider the special case of a nonlinear conservative system described by the boundary value problem

$$\frac{d^2 x(t)}{dt^2} + 1 + x(t)^3 = 0, \qquad x(0) = x(1) = 0.$$

In this situation, the vector $v_{\mathbf{x}}$ of (6.21) is given by

$$v_{\mathbf{x}} = (v_1, v_2, \ldots, v_8)^T, \qquad v_i = 1 + x_i^3, \qquad i = 1, 2, \ldots, 8, \tag{6.22}$$

if $m = 8$ is chosen. Next, we observe that a solution $\mathbf{x}^*$ of (6.21)–(6.22) satisfies

$$\|\mathbf{x}^*\| \leq \kappa^2 \|A^{-1}\| \, \|v_{\mathbf{x}^*}\| \qquad \Longrightarrow \qquad \|\mathbf{x}^*\| - \kappa^2 \|A^{-1}\| \left(1 + \|\mathbf{x}^*\|^3\right) \leq 0,$$

where $\|A^{-1}\| = 10$ and $\kappa = \frac{1}{9}$, so that $\|\mathbf{x}^*\| \in [0, \rho_1] \cup [\rho_2, +\infty]$, where $\rho_1 = 0.1236\ldots$ and $\rho_2 = 2.7821\ldots$ are the two positive roots of the scalar equation $81t - 10(1 + t^3) = 0$. In this case, we can consider

$$\mathbb{F} : \Lambda \subset \mathbb{R}^8 \longrightarrow \mathbb{R}^8 \qquad \text{with} \qquad \Lambda = \{\mathbf{x} \in \mathbb{R}^8 : \|\mathbf{x}\| < 2\},$$

since $\rho_1 < 2 < \rho_2$.

Moreover,

$$\mathbb{F}'(\mathbf{x}) = A + \kappa^2 \text{diag}(v'_{\mathbf{x}}), \qquad v'_{\mathbf{x}} = (v'_1, v'_2, \ldots, v'_8)^T, \qquad v'_i = 3x_i^2, \qquad i = 1, 2, \ldots, 8,$$

so that

$$\mathbb{F}'(\mathbf{x}) - \mathbb{F}'(\mathbf{y}) = 3\kappa^2 \text{diag}(\mathbf{z}),$$

Fig. 6.7 Domains of
parameters associated with the
Newton–Kantorovich
Theorem 2.1 (the black region)
and Theorem 6.4 when $\mu = \frac{3}{5}$
(black and pink regions) when
Newton's method is applied to
solve $\mathbb{F}(\mathbf{x}) = 0$ with $\mathbb{F}$ given in
(6.21) and $v_{\mathbf{x}}$ in (6.22)

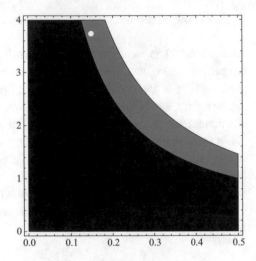

where $\mathbf{y} = (y_1, y_2, \ldots, y_8)^T$ and $\mathbf{z} = \left(x_1^2 - y_1^2, x_2^2 - y_2^2, \ldots, x_8^2 - y_8^2\right)^T$. In addition,

$$\|\mathbb{F}'(\mathbf{x}) - \mathbb{F}'(\mathbf{y})\| \le 3\kappa^2 \|\mathbf{x} + \mathbf{y}\| \|\mathbf{x} - \mathbf{y}\| \le 12\kappa^2 \|\mathbf{x} - \mathbf{y}\|,$$

$$\|\mathbb{F}'(\mathbf{x}) - \mathbb{F}'(\mathbf{x}_0)\| \le 3\kappa^2 \|\mathbf{x} + \mathbf{x}_0\| \|\mathbf{x} - \mathbf{x}_0\| \le 3\kappa^2 (2 + \|\mathbf{x}_0\|) \|\mathbf{x} - \mathbf{x}_0\|.$$

Thus, $L = \frac{4}{27}$ and $L_0 = \frac{1}{27}(2 + \|\mathbf{x}_0\|)$.

If we choose the starting point $\mathbf{x}_0 = (0.4, 0.4, \ldots, 0.4)^T$, we obtain $\beta = 10.5298\ldots$
and $\eta = 0.3549\ldots$, so that the condition of the Newton–Kantorovich theorem, $L\beta\eta \le \frac{1}{2}$,
is not satisfied, since $L\beta\eta = 0.5536\ldots > \frac{1}{2}$. Thus, Theorem 2.1 does not enable us to
apply Newton's method to approximate a solution of (6.21)–(6.22).

However, we can guarantee the convergence of Newton's method based on Theo-
rem 6.4, since $L_0 = \frac{4}{45}$, $\mu = \frac{3}{5}$, $\theta_L = 0.5536\ldots$ and condition $\theta_L \le \frac{(1-\mu\theta_L)^2}{\mu\theta_L} =$
$1.3423\ldots$ of Theorem 6.4 is then satisfied. In Fig. 6.7, we see graphically that $\mathbf{x}_0$ is
a good starting point for Newton's method, since its corresponding pair $(L, \beta\eta) =$
$\left(\frac{4}{27}, 3.7373\ldots\right)$, which is colored white, lies in the domain of parameters associated
with Theorem 6.4. So, we can now apply Newton's method to approximate a solution
of (6.21)–(6.22). We obtain the approximation given by the vector $\mathbf{x}^* = (x_1^*, x_2^*, \ldots, x_8^*)^T$
and shown in Table 6.1, which is reached after four iterations. Observe that $\|\mathbf{x}^*\| =$
$0.1236\ldots \le 2$, so that $\mathbf{x}^* \in \Lambda$. In Table 6.2 we show the errors $\|\mathbf{x}_n - \mathbf{x}^*\|$ and the
sequence $\{\|\mathbb{F}(\mathbf{x}_n)\|\}$. Notice that the vector shown in Table 6.1 is a good approximation of
the solution of system (6.21)–(6.22).

On the other hand, the convergence condition required in the result given by Argyros
and Hilout in [14] is also satisfied, so that the convergence of Newton's method is also
guaranteed by their result. In Table 6.3, we can see as the a priori error estimates $\{s_n\}$,

Table 6.1 Approximation of the solution $\mathbf{x}^*$ of (6.21)–(6.22)

i	x_i^*	i	x_i^*
1	0.049432...	5	0.123618...
2	0.086517...	6	0.111249...
3	0.111249...	7	0.086517...
4	0.123618...	8	0.049432...

Table 6.2 Absolute errors obtained by Newton's method and $\{\|\mathbb{F}(\mathbf{x}_n)\|\}$

n	$\|\mathbf{x}_n - \mathbf{x}^*\|$	$\|\mathbb{F}(\mathbf{x}_n)\|$
0	$3.5056\ldots \times 10^{-1}$	$3.8686\ldots \times 10^{-1}$
1	$1.0260\ldots \times 10^{-2}$	$1.3142\ldots \times 10^{-3}$
2	$3.2112\ldots \times 10^{-6}$	$4.5505\ldots \times 10^{-7}$
3	$3.1017\ldots \times 10^{-13}$	$4.7212\ldots \times 10^{-14}$

Table 6.3 A priori error estimates

n	s_n	$|t^* - t_n|$
0	$3.6654\ldots \times 10^{-1}$	$4.5369\ldots \times 10^{-1}$
1	$1.1621\ldots \times 10^{-2}$	$9.8795\ldots \times 10^{-2}$
2	$1.3557\ldots \times 10^{-3}$	$1.0530\ldots \times 10^{-2}$
3	$2.3264\ldots \times 10^{-5}$	$1.4616\ldots \times 10^{-4}$
4	$7.0893\ldots \times 10^{-9}$	$2.8952\ldots \times 10^{-8}$
5	–	$1.1364\ldots \times 10^{-15}$

obtained from (6.15) and the recurrence relation (II_n), defined in the proof of Theorem 6.4, improve slightly those of Argyros and Hilout $\{|t^* - t_n|\}$.

6.2 Operators with ω-Lipschitz Continuous First Derivative

We have seen in Chap. 3 that the semilocal convergence of Newton's method can be also studied under the following Hölder continuity condition for F':

(A2b)   There exist two constants $K \geq 0$ and $p \in (0, 1]$ such that $\|F'(x) - F'(y)\| \leq K\|x - y\|^p$, $x, y \in \Omega$.

We can also consider the case where F' comes from a combination of operators such that F' is Lipschitz or Hölder continuous in Ω, [30]. Thus, let us consider the condition

(A2c)   There exist $2m$ constants $K_1, K_2, \ldots, K_m \geq 0$ and $p_1, p_2, \ldots, p_m \in (0, 1]$ such that $\|F'(x) - F'(y)\| \leq \sum_{i=1}^{m} K_i \|x - y\|^{p_i}$, $x, y \in \Omega$.

To give sufficient generality to all the above conditions on the first Fréchet derivative F', we introduce the following condition:

(A2d)　There exist two continuous and nondecreasing functions $\omega : [0, +\infty) \rightarrow \mathbb{R}$ and $h : [0, 1] \rightarrow \mathbb{R}$ such that $\omega(0) \geq 0$, $\omega(tz) \leq h(t)\omega(z)$, with $t \in [0, 1]$, $z \in [0, \infty)$ and $\|F'(x) - F'(y)\| \leq \omega(\|x - y\|)$, $x, y \in \Omega$.

Obviously, we obtain (A2) when $\omega(z) = Lz$ and $h(t) = t$, (A2b) when $\omega(z) = Kz^p$ and $h(t) = t^p$, and (A2c) when $\omega(z) = \sum_{i=1}^{m} K_i z^{p_i}$ and $h(t) = t^p$, where $p = \min_{i=1,2,\ldots,m}\{p_i\}$ and $p_i \in (0, 1]$, for all $i = 1, 2, \ldots, m$. Moreover, we note that the condition $\omega(tz) \leq h(t)\omega(z)$, for $t \in [0, 1]$, is not restrictive, since there always exists a function h, such that $h(t) = 1$, for $t \in [0, 1]$, because ω is a nondecreasing function.

Accordingly, from now on, we focus on condition (A2d).

6.2.1　Convergence Analysis

Now, we present a semilocal convergence result for Newton's method under condition (A1) on the starting point x_0 and condition (A2d) on the first Fréchet derivative F'. First, we observe that, if (A2d) is satisfied, then once x_0 is fixed, there exist two continuous and nondecreasing functions $\omega_0 : [0, +\infty) \rightarrow \mathbb{R}$ and $h_0 : [0, 1] \rightarrow \mathbb{R}$ such that $\omega_0(0) \geq 0$, $\omega_0(tz) \leq h_0(t)\omega_0(z)$, for $t \in [0, 1]$, $z \in [0, \infty)$, and

$$\|F'(x) - F'(x_0)\| \leq \omega_0(\|x - x_0\|), \quad x \in \Omega, \tag{6.23}$$

where $\omega_0(z) \leq \omega(z)$, for $z \in [0, \infty)$. In particular, we can always take $\omega_0 = \omega$ and $h_0 = h$. In addition [38], we establish a semilocal convergence result by combining conditions (A2d) and (6.23). This allows us to enlarge the known domains of parameters, so that there are more choices of good starting points for Newton's method, as we will see later in applications.

6.2.1.1　Recurrence Relations

Suppose (A1)–(A2d) and denote $\theta_\omega = \beta\omega(\eta)$ and $\theta_\omega^0 = \beta\omega_0(\eta)$. We then consider $a_0 = \theta_\omega^0$, $b_0 = \theta_\omega$ and $d_0 = I_{h_0} a_0 f(a_0)$, where $I_{h_0} = \int_0^1 h_0(t)$ and

$$f(t) = \frac{1}{1-t}. \tag{6.24}$$

Next, we define $b_1 = b_0 f(a_0)h(d_0)$ and

$$b_n = b_{n-1} f(b_{n-1})h(e_{n-1}), \quad n \geq 2, \tag{6.25}$$

$$e_n = I_h b_n f(b_n), \quad n \geq 1, \tag{6.26}$$

where $I_h = \int_0^1 h(t)\,dt$.

Clearly, the interesting case is when $b_0 > 0$, since if $b_0 = 0$, the solution of the equation $F(x) = 0$ is x_0.

Now, let us prove the following four recurrence relations for sequences (6.1), (6.25), and (6.26):

$$\|\Gamma_1\| = \|[F'(x_1)]^{-1}\| \leq f(a_0)\|\Gamma_0\|, \tag{6.27}$$

$$\|x_2 - x_1\| \leq d_0\|x_1 - x_0\|, \tag{6.28}$$

$$\|\Gamma_1\|\omega(\|x_2 - x_1\|) \leq b_1, \tag{6.29}$$

$$\|x_2 - x_0\| \leq (1 + d_0)\|x_1 - x_0\|, \tag{6.30}$$

provided that

$$x_1 \in \Omega \quad \text{and} \quad a_0 < 1. \tag{6.31}$$

If $x_1 \in \Omega$, then

$$\|I - \Gamma_0 F'(x_1)\| \leq \|\Gamma_0\|\|F'(x_0) - F'(x_1)\| \leq \beta\omega_0(\eta) = a_0 < 1$$

and, by the Banach lemma on invertible operators, the operator Γ_1 exists and

$$\|\Gamma_1\| \leq \frac{\|\Gamma_0\|}{1 - \|I - \Gamma_0 F'(x_1)\|} < f(a_0)\|\Gamma_0\|.$$

Using a Taylor series expansion and the sequence (6.1), we see that

$$\|F(x_1)\| = \left\| \int_0^1 (F'(x_0 + \tau(x_1 - x_0)) - F'(x_0))(x_1 - x_0)\,d\tau \right\|$$

$$\leq \int_0^1 \omega_0(\tau\|x_1 - x_0\|)\,d\tau\,\|x_1 - x_0\|$$

$$\leq \left(\int_0^1 \omega_0(\tau\eta)\,d\tau \right)\|x_1 - x_0\|$$

$$\leq I_{h_0}\,\omega_0(\eta)\|x_1 - x_0\|.$$

Consequently,

$$\|x_2 - x_1\| \le \|\Gamma_1\| \|F(x_1)\| \le I_{h_0} a_0 f(a_0) \|x_1 - x_0\| = d_0 \|x_1 - x_0\|$$

$$\|\Gamma_1\| \omega(\|x_2 - x_1\|) \le \|\Gamma_0\| f(a_0) h(d_0) \omega(\|x_1 - x_0\|) \le b_0 f(a_0) h(d_0) = b_1,$$

$$\|x_2 - x_0\| \le \|x_2 - x_1\| + \|x_1 - x_0\| \le (1 + d_0)\|x_1 - x_0\| < \left(1 + \frac{d_0}{1 - e_1}\right) \eta = R,$$

provided that $d_0 \le 1$, namely $a_0 \le \frac{1}{1 + I_{h_0}}$, and $e_1 < 1$.

We also prove in the same way as above the following four recurrence relations for the sequences (6.1), (6.25), and (6.26):

$$\|\Gamma_2\| = \|[F'(x_2)]^{-1}\| \le f(b_1)\|\Gamma_1\|, \tag{6.32}$$

$$\|x_3 - x_2\| \le e_1\|x_2 - x_1\|, \tag{6.33}$$

$$\|\Gamma_2\| \omega(\|x_3 - x_2\|) \le b_2, \tag{6.34}$$

$$\|x_3 - x_0\| \le (1 + d_0(1 + e_1))\|x_1 - x_0\|, \tag{6.35}$$

provided that

$$x_2 \in \Omega \quad \text{and} \quad b_1 < 1. \tag{6.36}$$

In addition, we generalize the last recurrence relations to every point of sequence (6.1), so that we can guarantee that (6.1) is a Cauchy sequence from them. For this, we first analyze below the sequences $\{b_n\}$ and $\{e_n\}$.

6.2.1.2 Analysis of the Scalar Sequence

We analyze the scalar sequences defined in (6.25)–(6.26) in order to prove later the convergence of sequence (6.1). For this, it suffices to verify that (6.1) is a Cauchy sequence and (6.31)–(6.36) are true for all x_n and b_{n-1} with $n \ge 3$. We begin with a technical lemma.

Lemma 6.6 *Let f be the scalar function defined in (6.24). If $a_0 \le \frac{1}{1 + I_{h_0}}$ and b_1 is such that*

$$b_1 < \frac{1}{1 + I_h} \quad \text{and} \quad b_1 + h(e_1) < 1, \tag{6.37}$$

then

(a) *the sequences* $\{b_n\}$ *and* $\{e_n\}$ *are strictly decreasing;*
(b) $e_n < 1$ *and* $b_n < 1$, *for all* $n \geq 1$.

If $b_1 = 1 - h(e_1) < \frac{1}{1+I_h}$, *then* $e_n = e_1 < 1$ *and* $b_n = b_1 < 1$ *for all* $n \geq 2$.

Proof We first consider the case when b_1 satisfies (6.37). Item (a) is proved by induction on n. As $b_1 + h(e_1) < 1$, we have $b_2 < b_1$ and $e_2 < e_1$, since f is increasing. If we now assume that $b_j < b_{j-1}$ and $e_j < e_{j-1}$, for all $i = 2, 3, \ldots, n$, then

$$b_{n+1} = b_n f(b_n) h(e_n) < b_n f(b_1) h(e_1) < b_n,$$

$$e_{n+1} = I_h b_{n+1} f(b_{n+1}) < I_h b_n f(b_n) = e_n,$$

since f and h are nondecreasing functions in $[0, 1)$. Consequently, the sequences $\{e_n\}$ and $\{b_n\}$ are strictly decreasing for all $n \geq 2$.

To see item (b), we have $e_n < e_1 < 1$ and $b_n < b_1 < 1$, for all $n \geq 2$, by item (a) and the conditions given in (6.37).

Second, if $b_1 = 1 - h(e_1)$, then $h(e_1) f(b_1) = 1$ and so $b_n = b_1 = 1 - h(e_1) < 1$, for all $n \geq 2$. Moreover, if $b_1 < \frac{1}{1+I_h}$, then $e_n = e_1 < 1$, for all $n \geq 2$. ∎

6.2.1.3 Semilocal Convergence Result

We are now ready to prove the semilocal convergence of Newton's method under the conditions (A1)–(A2d) and

(A3b) $a_0 \leq \dfrac{1}{1 + I_{h_0}}$, b_1 satisfies (6.37) and $B(x_0, R) \subset \Omega$, where $R = \left(1 + \dfrac{d_0}{1-e_1}\right) \eta$.

Theorem 6.7 *Let* $F : \Omega \subseteq X \to Y$ *be a continuously Fréchet differentiable operator defined on a nonempty open convex domain* Ω *of a Banach space* X *with values in a Banach space* Y. *Suppose that the conditions* (A1)–(A2d)–(A3b) *are satisfied. Then, the Newton sequence* (6.1) *starting at* x_0 *converges to a solution* x^* *of the equation* $F(x) = 0$ *and* $x_n, x^* \in \overline{B(x_0, R)}$, *for all* $n = 0, 1, 2, \ldots$

Proof We begin by proving the following four statements for the sequences (6.1), (6.25) and (6.26), and $n \geq 3$:

(I_n)   There exists $\Gamma_{n-1} = [F'(x_{n-1})]^{-1}$ and $\|\Gamma_{n-1}\| \leq f(b_{n-2})\|\Gamma_{n-2}\|$;
(II_n) $\|x_n - x_{n-1}\| \leq e_{n-2}\|x_{n-1} - x_{n-2}\|$;
(III_n) $\|\Gamma_{n-1}\|\omega(\|x_n - x_{n-1}\|) \leq b_{n-1}$;
(IV_n) $x_n \in \Omega$.

Observe that $x_1 \in \Omega$, since $\eta < R$. Moreover, from (6.27)–(6.30), it follows that $x_2 \in \Omega$. Furthermore, (6.32)–(6.35) imply that items (I_3)–(II_3)–(III_3)–(IV_3) hold. If we now suppose that items (I_{n-1})–(II_{n-1})–(III_{n-1}) hold, then by analogy to the case (I_3)–(II_3)–(III_3) we conclude that items (I_n)–(II_n)–(III_n) also hold. Notice that $b_n < 1$ for all $n \geq 1$. Now, let us prove (IV_n). Item (II_n) implies that

$$\|x_n - x_0\| \leq \sum_{j=1}^{n} \|x_j - x_{j-1}\|$$

$$\leq \left(1 + \sum_{i=1}^{n-2} \left(\prod_{j=1}^{i} e_j\right)\right) \|x_2 - x_1\| + \|x_1 - x_0\|$$

$$< \left(1 + \sum_{i=1}^{n-2} e_1^i\right) \|x_2 - x_1\| + \|x_1 - x_0\|$$

$$< \frac{1}{1 - e_1} \|x_2 - x_1\| + \|x_1 - x_0\|$$

$$\leq \left(1 + \frac{d_0}{1 - e_1}\right) \|x_1 - x_0\|$$

$$\leq R,$$

so that $x_n \in B(x_0, R)$. As $B(x_0, R) \subset \Omega$, then $x_n \in \Omega$ for all $n \geq 0$. Note that the conditions given in (6.36) are satisfied for all x_n and b_{n-1} with $n \geq 3$.

Next, we prove that $\{x_n\}$ is a Cauchy sequence. For this, we proceed in much the same way as above. So, for $m \geq 2$ and $n \geq 2$, item (II_n) and Lemma 6.6 (a) imply that

$$\|x_{n+m} - x_n\| \leq \sum_{j=n}^{n+m-1} \|x_{j+1} - x_j\|$$

$$\leq \sum_{i=n-1}^{n+m-2} \left(\prod_{j=1}^{i} e_j\right) \|x_2 - x_1\|$$

$$< \left(\sum_{i=n-1}^{n+m-2} e_1^i\right) \|x_2 - x_1\|$$

$$= \left(\sum_{i=0}^{m-1} e_1^{n+i-1}\right) \|x_2 - x_1\|$$

$$= \frac{1 - e_1^m}{1 - e_1} e_1^{n-1} \|x_2 - x_1\|.$$

Thus, $\{x_n\}$ is a Cauchy sequence and then there exists $x^* \in \overline{B(x_0, R)}$ such that $x^* = \lim_n x_n$.

Let us verify that x^* is a solution of the equation $F(x) = 0$. Since $\|\Gamma_n F(x_n)\| \to 0$ as $n \to \infty$, if we take into account that

$$\|F(x_n)\| \leq \|F'(x_n)\| \|\Gamma_n F(x_n)\|$$

and $\{\|F'(x_n)\|\}$ is bounded, since

$$\|F'(x_n)\| \leq \|F'(x_0)\| + \omega_0(\|x_n - x_0\|) \leq \|F'(x_0)\| + \omega_0(R),$$

we conclude that $\|F(x_n)\| \to 0$ as $n \to \infty$. Therefore, $F(x^*) = 0$ thanks to the continuity of the operator F in $\overline{B(x_0, R)}$. ∎

6.2.2 Domain of Parameters and Applications

The main aim of this section is to improve the domain of starting points for Newton's method under conditions (A1)–(A2d)–(A3b). For this, with the help of center conditions of type (6.23), we introduce new parameters associated with the starting point x_0, which allows us to relax the conditions required for x_0. We do this based on an analysis of the domain of parameters.

On the other hand, using a discretization of a special kind of nonlinear conservative problems, we obtain a system of nonlinear equations that we use as a test to demonstrate the improvement obtained, i.e., the extension of the domain of starting points for Newton's method.

6.2.2.1 Domain of Parameters

The aim of this section is to consider the general semilocal convergence conditions given in (A1)–(A2c)–(A3b) for the starting point x_0 and determine when the semilocal convergence of Newton's method is guaranteed from x_0.

Observe that the previous conditions associated with x_0 in general involve positive parameters $K_1, K_2, \ldots, K_m, \beta, \eta$ and $p_1, p_2, \ldots, p_m \subset (0, 1]$. Moreover, we consider the condition corresponding to (6.23) for this case, namely,

$$\|F'(x) - F'(x_0)\| \leq \sum_{i=1}^{m} K_i^0 \|x - x_0\|^{p_i}. \tag{6.38}$$

Thus, new positive real parameters $K_1^0, K_2^0, \ldots, K_m^0$ appear. Obviously, $K_i^0 \leq K_i$, for all $i = 1, 2, \ldots, m$. With the new parameters $K_1^0, K_2^0, \ldots, K_m^0$, which are not commonly used, we can relax the conditions on x_0.

Next, we adapt Theorem 6.7 to the present setting. For this, we set

$$\omega(z) = \sum_{i=1}^{m} K_i z^{p_i} \le K \sum_{i=1}^{m} z^{p_i}, \quad \text{where} \quad K = \max_{i=1,2,\dots,m} \{K_i\},$$

and, for $t \in [0, 1]$, we observe that $\omega(tz) \le t^p \omega(z)$, where $p = \min_{i=1,2,\dots,m}\{p_i\}$, so that $h(t) = t^p$.

On the other hand, to treat the condition (6.38), we can define $\omega_0 : [0, +\infty) \to \mathbb{R}$ and $h_0 : [0, 1] \to \mathbb{R}$, such that $\omega_0(0) \ge 0$, $\omega_0(tz) \le h_0(t)\omega_0(z)$, with $t \in [0, 1]$ and $z \in [0, \infty)$, and $\|F'(x) - F'(x_0)\| \le \omega_0(\|x - x_0\|)$, with $x, x_0 \in \Omega$. Specifically, we take

$$\omega_0(z) = \sum_{i=1}^{m} K_i^0 z^{p_i} \le K_0 \sum_{i=1}^{m} z^{p_i}, \quad \text{where} \quad K_0 = \max_{i=1,2,\dots,m} \{K_i^0\},$$

and, for $t \in [0, 1]$, we observe that $\omega_0(tz) \le t^p \omega_0(z)$, where $p = \min_{i=1,2,\dots,m}\{p_i\}$, so that $h_0(t) = t^p$.

So, exploiting the variation of parameters $K_1, K_2, \dots, K_m, K_1^0, K_2^0, \dots, K_m^0, \beta$ and η, we obtain a semilocal convergence result for Newton's method.

En route, we will see that the conditions on x_0 are relaxed by introducing the condition (6.38) and, with it, the new parameters.

To guarantee the convergence of Newton's method based on Theorem 6.7, the following three conditions must be satisfied:

$$a_0 \le \frac{1+p}{2+p}, \qquad b_1 < \frac{1+p}{2+p} \qquad \text{and} \qquad b_1 + \frac{b_1^p}{(1+p)^p(1-b_1)^p} < 1, \qquad (6.39)$$

where $p \in (0, 1]$. Using the auxiliary function

$$\phi(t; p) = (1+p)^p(1-t)^{1+p} - t^p, \quad p \in (0, 1], \qquad (6.40)$$

the third condition of (6.39) can be written as

$$\phi(b_1; p) > 0. \qquad (6.41)$$

Observe that $\phi(x; p)$ is nonincreasing and conve, $\phi(0; p) = (1+p)^p > 0$ and $\phi\left(\frac{1}{2}; p\right) \le 0$, for all $p \in (0, 1]$. Moreover, if $p = 1$, the unique zero of $\phi(t; 1)$ in the interval $\left(0, \frac{1}{2}\right]$ is $\frac{1}{2}$. If we now denote, for a fixed $p \in (0, 1]$, the unique zero of $\phi(t; p)$ in $\left(0, \frac{1}{2}\right]$ by $\xi(p)$ and require that $b_1 < \xi(p)$, then condition (6.41) holds. Moreover, since $\xi(p) \le \frac{1}{2}$, the second condition in (6.39) also holds.

Since

$$b_1 = b_0 f(a_0) h(d_0) = \frac{a_0^p b_0}{(1+p)^p (1-a_0)^{1+p}},$$

the second and third conditions in (6.39) are satisfied, provided that

$$b_0 < \xi(p) \frac{(1+p)^p (1-a_0)^{1+p}}{a_0^p}.$$

Now let write these conditions in terms of the initial parameters associated with the starting point x_0. So,

$$a_0 = \beta \omega_0(\eta) = \beta \sum_{i=1}^{m} K_i^0 \eta^{p_i} \leq \beta K_0 \sum_{i=1}^{m} \eta^{p_i} = \mu \beta K \sum_{i=1}^{m} \eta^{p_i} = \mu K \beta s(\eta),$$

where $K = \max_{i=1,2,\ldots,m}\{K_i\}$, $K_0 = \max_{i=1,2,\ldots,m}\{K_i^0\}$, $\mu = \frac{K_0}{K} \in (0,1]$ and $s(\eta) = \sum_{i=1}^{m} \eta^{p_i}$.

As in this case, we are dealing with sums of Lipschitz continuous and Hölder continuous operators, we denote $\theta_S = K\beta s(\eta)$ and $\theta_S^0 = K_0 \beta s(\eta)$; therefore, $\theta_S^0 = \mu \theta_S$, as in the previous cases. The first condition in (6.39) now reads

$$\theta_S = K\beta s(\eta) \leq \frac{1+p}{\mu(2+p)},$$

and the second and third conditions in (6.39) are satisfied if

$$\theta_S < \xi(p) \frac{(1+p)^p (1-\mu\theta_S)^{1+p}}{\mu^p \theta_S^p},$$

since $b_0 = \theta_S$ and $a_0 = \theta_S^0 = \mu\theta_S$. In addition, we can give the following result.

Corollary 6.8 *Let $F : \Omega \subseteq X \to Y$ be a continuously Fréchet differentiable operator defined on a nonempty open convex domain Ω of a Banach space X with values in a Banach space Y. Suppose that the conditions (A1)–(A2c) are satisfied. If*

$$\theta_S \leq \frac{1+p}{\mu(2+p)}, \qquad \theta_S < \xi(p) \frac{(1+p)^p (1-\mu\theta_S)^{1+p}}{\mu^p \theta_S^p},$$

where $p = \min_{i=1,2,\ldots,m}\{p_i\}$ and $\xi(p)$ is the unique zero of the function (6.40) in the interval $\left(0, \frac{1}{2}\right]$, and $B(x_0, R) \subset \Omega$, where $R = \left(1 + \frac{\upsilon}{1-\upsilon}\right)\eta$, $\upsilon = \frac{\mu\theta_S}{(1+p)(1-\mu\theta_S)}$, $\nu =$

$\frac{\vartheta}{(1+p)(1-\vartheta)}$ and $\vartheta = \frac{\theta_S \upsilon^p}{1-\mu\theta_S}$, then the Newton sequence (6.1) starting at x_0 converges to a solution x^* of the equation $F(x) = 0$ and $x_n, x^* \in \overline{B(x_0, R)}$, for all $n = 0, 1, 2, \ldots$

We can then define the domains of parameters

$$D_{GC}^{\mu} = \left\{ (x, y) \in \mathbb{R}_+^2 : xy \leq \frac{1+p}{\mu(2+p)}, \ xy < \xi(p) \frac{(1+p)^p (1 - \mu xy)^{1+p}}{\mu^p x^p y^p} \right\},$$

for each $\mu = K_0/K \in (0, 1]$, $p \in (0, 1]$, where $\xi(p)$ is the unique zero of the function (6.40) in the interval $\left(0, \frac{1}{2}\right]$.

Then, we define the parameters K, β and $s(\eta)$ for the starting point x_0, which satisfies (A1); Newton's method is convergent if there exists $\mu = K_0/K \in (0, 1]$ such that $(K, \beta s(\eta)) \in D_{GC}^{\mu}$.

Next, we compare the semilocal convergence conditions in the last corollary with those that appear in the following semilocal convergence result given by Ezquerro and Hernández in [28], where the technique of recurrence relations is also used to prove the convergence of Newton's method.

Theorem 6.9 (Theorem 2.2 of [28]) *Let $F : \Omega \subseteq X \to Y$ be a continuously Fréchet differentiable operator defined on a nonempty open convex domain Ω of a Banach space X with values in a Banach space Y. Suppose that the conditions (A1)-(A2c) are satisfied. If*

$$\theta_S \leq \frac{1+p}{2+p} \quad \text{and} \quad \theta_S < 1 - \frac{\theta_S^p}{(1+p)^p (1 - \theta_S)^p},$$

where $\theta_S = K\beta s(\eta)$ and $p = \min_{i=1,2,\ldots,m}\{p_i\}$, and $B(x_0, R) \subset \Omega$, where $R = \frac{(1+p)(1-\theta_S)\eta}{1+p-(2+p)\theta_S}$, then the Newton sequence (6.1) starting at x_0 converges to a solution x^* of the equation $F(x) = 0$ and $x_n, x^* \in \overline{B(x_0, R)}$, for all $n = 0, 1, 2, \ldots$

The corresponding domain of parameters

$$D_{EH} = \left\{ (x, y) \in \mathbb{R}_+^2 : xy \leq \frac{1+p}{2+p}, \ xy < 1 - \frac{x^p y^p}{(1+p)^p (1 - xy)^p} \right\},$$

where $p \in (0, 1]$.

Let us compare the semilocal convergence conditions in Corollary 6.8 and Theorem 6.9 to show that the domain of starting points given by Theorem 6.9 can be enlarged by using Corollary 6.8. We depict the domains of parameters associated with the two results in Fig. 6.8 which shows that the smaller the quantity $\mu = K_0/K \in (0, 1]$, the larger the domain of parameters: orange for $\mu = 0.2$, green for $\mu = 0.4$, red for $\mu = 0.6$ and yellow

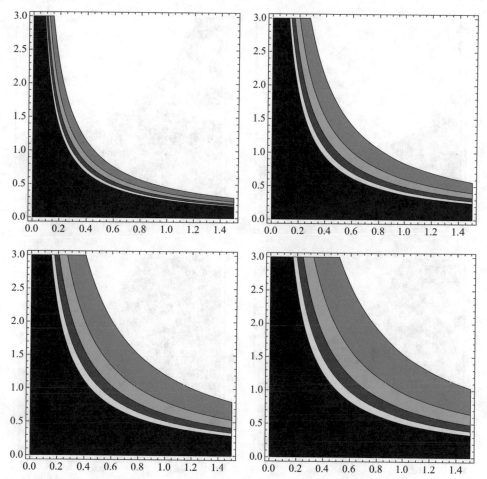

Fig. 6.8 Domains of parameters of Newton's method associated with Theorem 6.9 (the black region) and Corollary 6.8 for $p = \frac{1}{4}, \frac{1}{2}, \frac{3}{4}, 1$

for $\mu = 0.8$. Note that, as $\mu \to 1$, the domain of parameters provided by Corollary 6.8, D_{GC}^{μ}, tends to that provided by Theorem 6.9 (the black region). As a consequence,

$$D_{EH} = D_{GC}^1 = D_{GC}^{\mu_j} \subset D_{GC}^{\mu_{j-1}} \subset \cdots \subset D_{GC}^{\mu_0} \quad \text{for} \quad \mu_0 < \cdots < \mu_{j-1} < \mu_j = 1.$$

On the other hand, Fig. 6.9 makes clear the relationship between the domains of parameters associated with Corollary 6.8 and Theorem 6.9 with the variation of μ. For this, we placed the values of μ on the x-axis and the values of $\theta_S = K\beta s(\eta)$ in the y-axis. The aim of Fig. 6.9 is to see that the domain of parameters associated with Corollary 6.8 (blue and gray regions) is always larger, for all $\mu \in (0, 1]$, than that associated with Theorem 6.9 (the gray region).

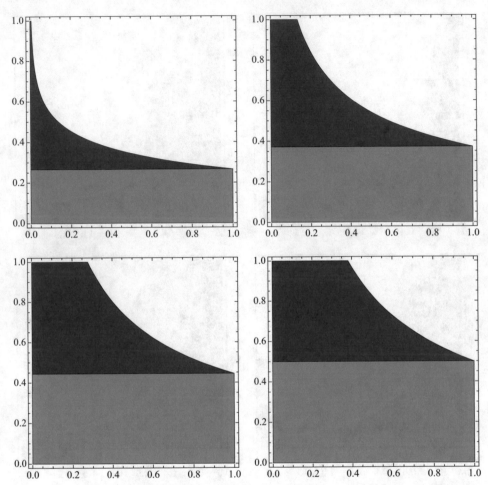

Fig. 6.9 Domains of parameters of Newton's method associated with Theorem 6.9 (the gray region) and Corollary 6.8 for $p = \frac{1}{4}, \frac{1}{2}, \frac{3}{4}, 1$ (blue and gray regions)

Let us prove analytically what the graphically evident inclusion, $D_{\mathrm{EH}} \subset D_{\mathrm{GC}}^{\mu}$, for each $\mu \in (0, 1]$. First, if $(K, \beta s(\eta)) \in D_{\mathrm{EH}}$, then $\theta_{\mathrm{S}} \leq \frac{1+p}{2+p}$ and, consequently, $\theta_{\mathrm{S}} \leq \frac{1+p}{\mu(2+p)}$, since $\mu \in (0, 1]$. Second, as $\theta_{\mathrm{S}} < 1 - \frac{\theta_{\mathrm{S}}^{p}}{(1+p)^{p}(1-\theta_{\mathrm{S}})^{p}}$, or equivalently $\phi(\theta_{\mathrm{S}}; p) > 0$, it follows, in view of the features of the function $\phi(x; p)$, that $\theta_{\mathrm{S}} < \xi(p)$ and $\mu\theta_{\mathrm{S}} < \xi(p)$, for each $\mu \in (0, 1]$, so that

$$\frac{(1 + p)^{p}(1 - \mu\theta_{\mathrm{S}})^{1+p}}{\mu^{p}\theta_{\mathrm{S}}^{p}} > 1,$$

since $\phi(\mu\theta_S; p) > 0$. Therefore, we have

$$\theta_S < \xi(p) \le \xi(p)\frac{(1+p)^p(1-\mu\theta_S)^{1+p}}{\mu^p\theta_S^p}.$$

Consequently, $(K, \beta s(\eta)) \in D_{GC}^\mu$ for each $\mu \in (0, 1]$.

Finally, let us show that $D_{GC}^{\mu_2} \subset D_{GC}^{\mu_1}$ if $\mu_1 < \mu_2$ with $\mu_1, \mu_2 \in (0, 1]$. Since $\psi(\mu) = \frac{1+p}{(2+p)\mu}$ and $\varphi(\mu) = \xi(p)\frac{(1+p)^p(1-\mu\theta_S)^{1+p}}{\mu^p\theta_S^p}$, we have $\psi'(\mu) = \frac{-(1+p)}{(2+p)\mu^2} \le 0$ and $\varphi'(\mu) = -\frac{(1+p)^p\xi(p)(1-\mu\theta_S)^p(p+\mu\theta_S)}{\mu^{1+p}\theta_S^p} \le 0$, because $1-\mu\theta_S \ge \frac{1}{2+p} \ge 0$. Therefore, $\psi(\mu_2) \le \psi(\mu_1)$ and $\varphi(\mu_2) \le \varphi(\mu_1)$, and therefore $D_{GC}^{\mu_2} \subset D_{GC}^{\mu_1}$.

To that end, we have proved the improvement obtained for the domain of parameters of Newton's method with the help of conditions of type (6.23) that we have just shown in Fig. 6.9.

6.2.2.2   Application to a Conservative Problem

Next, we apply the semilocal convergence result obtained above to a boundary value problem of type (6.18)–(6.19).

Example 6.10   Consider the law $\Psi(x(t)) = 1 + x(t)^{\frac{5}{3}} + x(t)^3$ for heat generation. Then, the vector $v_\mathbf{x}$ of (6.21) with $\mathfrak{m} = 8$ is

$$v_\mathbf{x} = (v_1, v_2, \ldots, v_8)^t, \qquad v_i = 1 + x_i^{\frac{5}{3}} + x_i^3, \qquad i = 1, 2, \ldots, 8. \tag{6.42}$$

In this case, we observe that a solution $\mathbf{x}^*$ of problem (6.21)–(6.42) satisfies

$$\|\mathbf{x}^*\| \le \kappa^2\|A^{-1}\|\,\|v_{\mathbf{x}^*}\| \quad\Longrightarrow\quad \|\mathbf{x}^*\| - \kappa^2\|A^{-1}\|\left(1 + \|\mathbf{x}^*\|^{\frac{5}{3}} + \|\mathbf{x}^*\|^3\right) \le 0,$$

where $\|A^{-1}\| = 10$ and $\kappa = \frac{1}{9}$, so that $\|\mathbf{x}^*\| \in [0, \rho_1] \cup [\rho_2, +\infty)$, where $\rho_1 = 0.1277\ldots$ and $\rho_2 = 2.4253\ldots$ are the two positive roots of the scalar equation $81t - 10\left(1 + t^{\frac{5}{3}} + t^3\right) = 0$. In this case, we can consider

$$\mathbb{F} : \Lambda \subset \mathbb{R}^8 \longrightarrow \mathbb{R}^8 \qquad \text{with} \qquad \Lambda = \left\{\mathbf{x} \in \mathbb{R}^8 : \|\mathbf{x}\| < \frac{3}{4}\right\},$$

since $\rho_1 < \frac{3}{4} < \rho_2$.

Then

$$\mathbb{F}'(\mathbf{x}) = A + \kappa^2\mathrm{diag}(v'_\mathbf{x}), \quad v'_\mathbf{x} = (v'_1, v'_2, \ldots, v'_8)^t, \quad v'_i = \frac{5}{3}x_i^{\frac{2}{3}} + 3x_i^2, \quad i = 1, 2, \ldots, 8,$$

and

$$\mathbb{F}'(\mathbf{x}) - \mathbb{F}'(\mathbf{y}) = \kappa^2 \left(\frac{5}{3} \operatorname{diag}(\mathbf{z}) + 3 \operatorname{diag}(\mathbf{w}) \right),$$

where $\mathbf{y} = (y_1, y_2, \ldots, y_8)^T$, $\mathbf{z} = \left(x_1^{\frac{2}{3}} - y_1^{\frac{2}{3}}, x_2^{\frac{2}{3}} - y_2^{\frac{2}{3}}, \ldots, x_8^{\frac{2}{3}} - y_8^{\frac{2}{3}} \right)^T$ and $\mathbf{w} = \left(x_1^2 - y_1^2, x_2^2 - y_2^2, \ldots, x_8^2 - y_8^2 \right)^T$. Further,

$$\|\mathbb{F}'(\mathbf{x}) - \mathbb{F}'(\mathbf{y})\| \leq \kappa^2 \left(\frac{5}{3} \left\| \mathbf{x}^{\frac{1}{3}} + \mathbf{y}^{\frac{1}{3}} \right\| \left\| \mathbf{x}^{\frac{1}{3}} - \mathbf{y}^{\frac{1}{3}} \right\| + 3 \|\mathbf{x} + \mathbf{y}\| \|\mathbf{x} - \mathbf{y}\| \right)$$

$$\leq \kappa^2 \left(\frac{10}{3} \left(\frac{3}{4} \right)^{\frac{1}{3}} \|\mathbf{x} - \mathbf{y}\|^{\frac{1}{3}} + \frac{9}{2} \|\mathbf{x} - \mathbf{y}\| \right)$$

$$\leq \frac{9}{2} \kappa^2 \left(\|\mathbf{x} - \mathbf{y}\|^{\frac{1}{3}} + \|\mathbf{x} - \mathbf{y}\| \right),$$

$$\|\mathbb{F}'(\mathbf{x}) - \mathbb{F}'(\mathbf{x}_0)\| \leq \kappa^2 \left(\frac{5}{3} \left(\|\mathbf{x}\|^{\frac{1}{3}} + \|\mathbf{x}_0\|^{\frac{1}{3}} \right) \|\mathbf{x} - \mathbf{x}_0\|^{\frac{1}{3}} + 3 \left(\|\mathbf{x}\| + \|\mathbf{x}_0\| \right) \|\mathbf{x} - \mathbf{x}_0\| \right)$$

$$\leq \kappa^2 \max \left\{ \frac{5}{3} \left(\left(\frac{3}{4} \right)^{\frac{1}{3}} + \|\mathbf{x}_0\|^{\frac{1}{3}} \right), 3 \left(\frac{3}{4} + \|\mathbf{x}_0\| \right) \right\} \left(\|\mathbf{x} - \mathbf{x}_0\|^{\frac{1}{3}} + \|\mathbf{x} - \mathbf{x}_0\| \right).$$

Thus, $K = \frac{1}{18} = 0.0555\ldots$, $K_0 = \frac{1}{81} \max \left\{ \frac{5}{3} \left(\left(\frac{3}{4} \right)^{\frac{1}{3}} + \|\mathbf{x}_0\|^{\frac{1}{3}} \right), 3 \left(\frac{3}{4} + \|\mathbf{x}_0\| \right) \right\}$ and $p = \frac{1}{3}$.

If we choose the starting point $\mathbf{x}_0 = (0, 0, \ldots, 0)^T$, we obtain $\beta = 10$ and $\eta = 0.1234\ldots$, so the condition $\theta_S < 1 - \frac{\theta_S^p}{(1+p)^p (1-\theta_S)^p}$ of Theorem 6.9 is not satisfied, since $\theta_S = 0.3452\ldots > 1 - \frac{\theta_S^p}{(1+p)^p (1-\theta_S)^p} = 0.2660\ldots$ Hence, we cannot use Theorem 6.9 to guarantee the convergence of Newton's method for approximating a solution of (6.21)–(6.42).

However, we can guarantee convergence based on from Corollary 6.8, since $K_0 = \frac{1}{36} = 0.0277\ldots$, $\mu = \frac{1}{2}$ and so the conditions $\theta_S \leq \frac{1+p}{\mu(2+p)} = \frac{8}{7}$ and $\theta_S < \xi(p) \frac{(1+p)^p (1-\mu\theta_S)^{1+p}}{\mu^p \theta_S^p} = 0.4716\ldots$ of Corollary 6.8 are satisfied. In Fig. 6.10, we see graphically that $\mathbf{x}_0$ is a good starting point for applying Newton's method, since the corresponding pair $(K, \beta s(\eta)) = \left(\frac{1}{18}, 6.2139\ldots \right)$, colored white, is in the domain of parameters associated with Corollary 6.8. Applying Newton's method, we obtain the approximation given by $\mathbf{x}^* = (x_1^*, x_2^*, \ldots, x_8^*)^T$ and shown in Table 6.4, which is reached after four iterations. Observe that $\|\mathbf{x}^*\| = 0.1267\ldots \leq \frac{3}{4}$, so that $\mathbf{x}^* \in \Lambda$. In Table 6.5 we show the errors $\|\mathbf{x}_n - \mathbf{x}^*\|$ and the sequence $\{\|\mathbb{F}(\mathbf{x}_n)\|\}$. Notice that the vector shown in Table 6.4 is a good approximation of the solution of system (6.21)–(6.42).

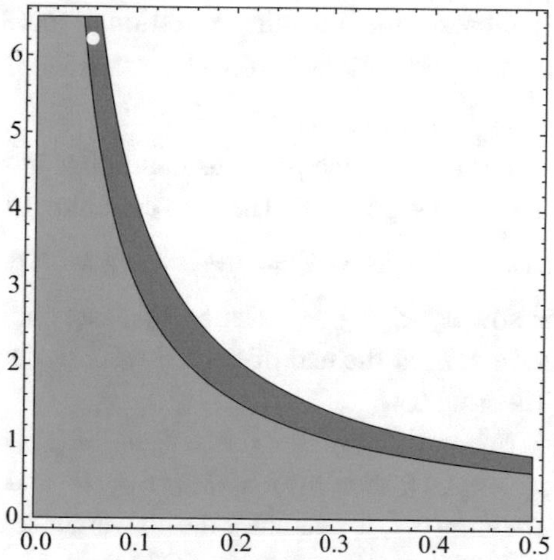

Fig. 6.10 Domains of parameters associated with Theorem 6.9 (the green region) and Corollary 6.8 when $\mu = \frac{1}{2}$ (pink and green regions) and Newton's method is applied to solve the equation $\mathbb{F}(\mathbf{x}) = 0$ with $\mathbb{F}$ given in (6.21) and $v_{\mathbf{x}}$ in (6.42)

Table 6.4 Approximation of the solution $\mathbf{x}^*$ of (6.21)–(6.42)

i	x_i^*	i	x_i^*
1	0.050463...	5	0.126720...
2	0.088494...	6	0.113955...
3	0.113955...	7	0.088494...
4	0.126720...	8	0.050463...

Table 6.5 Absolute errors obtained by Newton's method and $\{\|\mathbb{F}(\mathbf{x}_n)\|\}$

n	$\|\mathbf{x}_n - \mathbf{x}^*\|$	$\|\mathbb{F}(\mathbf{x}_n)\|$
0	$1.2672\ldots \times 10^{-1}$	$1.2345\ldots \times 10^{-2}$
1	$3.2637\ldots \times 10^{-3}$	$4.0112\ldots \times 10^{-4}$
2	$1.4751\ldots \times 10^{-6}$	$1.9526\ldots \times 10^{-7}$
3	$2.9716\ldots \times 10^{-13}$	$3.9927\ldots \times 10^{-14}$

6.2.2.3 Particular Cases

Observe that the Lipschitz and Hölder cases are particular cases of the one discussed above.

First, if $\omega_0(z) = L_0 z$ and $\omega(z) = Lz$, where L_0 and L are constants, then $h(t) = h_0(t) = t$ and F' is Lipschitz continuous in Ω. According to previously seen, to guarantee the convergence of Newton's method based on Theorem 6.7, the following three conditions must be satisfied:

$$a_0 \leq \frac{2}{3}, \qquad b_1 \leq \frac{2}{3} \qquad \text{and} \qquad b_1 + \frac{b_1}{2(1 - b_1)} < 1. \tag{6.43}$$

Observe that the third condition in (6.43) is satisfied if $b_1 < \frac{1}{2}$ and so the second and the third conditions of (6.43) are satisfied if $b_1 < \frac{1}{2}$. Moreover, as $b_1 = \frac{a_0 b_0}{2(1-a_0)^2}$, we have $b_1 < \frac{1}{2}$ provided that $b_0 < \frac{(1-a_0)^2}{a_0}$.

Let us write the previous conditions in terms of the initial parameters associated with the starting point x_0. The first condition of (6.43) reads $b_0 \leq \frac{2}{3\mu}$, since in the Lipschitz case $a_0 = L\beta\eta = \theta_L^0 = \mu\theta_L = \mu b_0$ with $\mu = \frac{L_0}{L} \in (0, 1]$ and the condition $b_0 < \frac{(1-a_0)^2}{a_0}$ is now $\theta_L < \frac{(1-\mu\theta_L)^2}{\mu\theta_L}$. Hence, the last condition is then satisfied if $\theta_L < \frac{\sqrt{\mu}-\mu}{\mu(1-\mu)}$, as we are have seen at the end of Sect. 6.1.1.3. As a consequence, we obtain the same result as in Theorem 6.4.

Second, if $\omega_0(z) = K_0 z^p$ and $\omega(z) = K z^p$, where K_0 and K are constants and $p \in (0, 1]$, then $h(t) = h_0(t) = t^p$ and F' is Hölder continuous in Ω. According to Theorem 6.7, to guarantee the convergence of Newton's method, the three conditions given in (6.39) must be satisfied with $p \in (0, 1]$.

Arguing analogously to how we did in Sect. 6.2.2.1, we see that these three conditions reduce to

$$b_0 = K\beta\eta^p \leq \frac{1+p}{\mu(2+p)} \qquad \text{and} \qquad b_0 \leq \xi(p)\frac{(1+p)^p(1-\mu b_0)^{1+p}}{\mu^p b_0^p},$$

where $\mu = K_0/K \in (0, 1]$, $p \in (0, 1]$ and $\xi(p)$ is the unique zero of the function (6.40) in the interval $\left(0, \frac{1}{2}\right]$. In this case, as F' is Hölder continuous, we consider $\theta_H = K\beta\eta^p$ and $\theta_H^0 = K_0\beta\eta^p = \mu\theta_H$ where $\mu = K_0/K \in (0, 1]$. In addition, we give the following result.

Corollary 6.11 (Corollary 4 of [33]) *Let $F : \Omega \subseteq X \to Y$ be a continuously Fréchet differentiable operator defined on a nonempty open convex domain Ω of a Banach space X with values in a Banach space Y. Suppose that the conditions (A1)–(A2b) are satisfied. If*

$$\theta_H \leq \frac{1+p}{\mu(2+p)}, \qquad \theta_H \leq \xi(p)\frac{(1+p)^p(1-\mu\theta_H)^{1+p}}{\mu^p \theta_H^p}, \tag{6.44}$$

with $\mu = K_0/K \in (0, 1]$, $p \in (0, 1]$ and if $\xi(p)$ the unique zero of the function (6.40) in the interval $\left(0, \frac{1}{2}\right]$, and $B(x_0, R) \subset \Omega$, where $R = \left(1 + \frac{\upsilon}{1-\upsilon}\right)\eta$, $\upsilon = \frac{\mu\theta_H}{(1+p)(1-\mu\theta_H)}$, $v = \frac{\vartheta}{(1+p)(1-\vartheta)}$ and $\vartheta = \frac{\theta_H\upsilon^p}{1-\mu\theta_H}$, then the Newton sequence (6.1) starting at x_0 converges to a solution x^* of the equation $F(x) = 0$ and $x_n, x^* \in \overline{B(x_0, R)}$, for all $n = 0, 1, 2, \ldots$

The domain of parameters associated with this corollary is

$$D_{\text{Höl}}^{\mu} = \left\{ (x, y) \in \mathbb{R}_+^2 : xy \leq \frac{1+p}{\mu(2+p)}, \ xy \leq \xi(p) \frac{(1+p)^p(1-\mu xy)^{1+p}}{\mu^p x^p y^p} \right\},$$

where $\mu = K_0/K \in (0, 1]$, $p \in (0, 1]$ and $\xi(p)$ is the unique zero of the function (6.40) in the interval $\left(0, \frac{1}{2}\right]$.

Notice that if $p = 1$, Corollary 6.11 reduces to Theorem 6.4.

Next, we compare the conditions required for the semilocal convergence of Newton's method in the last corollary with those that appear in the following semilocal convergence results obtained by Hernández in [48] and Keller in [58]:

Theorem 6.12 (Theorem 2.1 of [48]) *Let $F : \Omega \subseteq X \to Y$ be a continuously Fréchet differentiable operator defined on a nonempty open convex domain Ω of a Banach space X with values in a Banach space Y. Suppose that the (A1)–(A2b) are satisfied. If $\theta_H = K\beta\eta^p \leq \xi(p)$, with $p \in (0, 1]$ and if $\xi(p)$ is the unique zero of the function (6.40) in the interval $\left(0, \frac{1}{2}\right]$, and $B(x_0, R) \subset \Omega$, where $R = \frac{(1+p)(1-\theta_H)}{(1+p)-(2+p)\theta_H}\eta$, then the Newton sequence (6.1) starting at x_0 converges to a solution x^* of the equation $F(x) = 0$ and $x_n, x^* \in \overline{B(x_0, R)}$, for all $n = 0, 1, 2, \ldots$*

For each fixed p, the domain of parameters that we can associated with this result is

$$D_H = \left\{ (x, y) \in \mathbb{R}_+^2 : xy \leq \xi(p) \right\},$$

where $p \in (0, 1]$ and $\xi(p)$ is the unique zero of the function (6.40) in the interval $\left(0, \frac{1}{2}\right]$.

Theorem 6.13 (Theorem 4 of [58]) *Let $F : \Omega \subseteq X \to Y$ be a continuously Fréchet differentiable operator defined on a nonempty open convex domain Ω of a Banach space X with values in a Banach space Y. Suppose that the (A1)–(A2b) are satisfied and that $\theta_H \leq \frac{1}{2+p} \left(\frac{p}{1+p}\right)^p$, with $p \in (0, 1]$. Let $R_0(p)$ be the smallest positive root of $(2+p)K\beta t^{1+p} - (1+p)(t-\eta) = 0$. If $\varrho \geq R_0(p)$, then the equation $F(x) = 0$ has a solution $x^* \in \overline{B(x_0, \varrho)}$ and the Newton sequence converges to this solution with initial point x_0.*

For each fixed p, the domain of parameters that we can associate with this theorem is

$$D_K = \left\{ (x, y) \in \mathbb{R}_+^2 : xy \leq \frac{1}{2+p} \left(\frac{p}{1+p}\right)^p \right\},$$

where $p \in (0, 1]$.

Let us compare the conditions required for the semilocal convergence of Newton's method in the last three results, Corollary 6.11 and Theorems 6.12 and 6.13, and

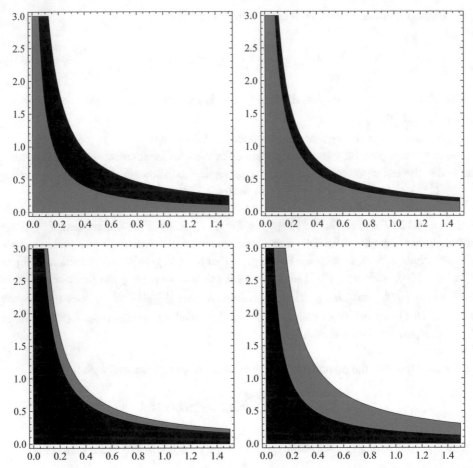

Fig. 6.11 Domains of parameters of Newton's method associated with Theorem 6.12 (the pink region) and Theorem 6.13 (the black region) for $p = \frac{1}{10}, \frac{1}{5}, \frac{2}{5}, \frac{4}{5}$

demonstrate that based on Corollary 6.11 we can increase the domain of starting points given by Theorems 6.12 and 6.13.

First, we note that Hernández proves in [48] that $D_K \subset D_H$ if $p \in (0.2856\ldots, 1]$ and $D_H \subset D_K$ if $p \in [0, 0.2856\ldots]$, as shown in Fig. 6.11, for $p = \frac{1}{10}, \frac{1}{5}, \frac{2}{5}, \frac{4}{5}$.

Second, Fig. 6.12 shows that $D_H \subset D_{\text{Höl}}^{\mu}$ for, $p = \frac{1}{10}, \frac{1}{5}, \frac{2}{5}, \frac{4}{5}$, so that we can guess that the smaller the quantity $\mu = \frac{K_0}{K} \in (0, 1]$, the larger the domain of parameters: orange for $\mu = 0.1$, green for $\mu = 0.2$, red for $\mu = 0.4$ and yellow for $\mu = 0.8$. Note that, as $\mu \to 1$, the domain of parameters associated with Corollary 6.11, $D_{\text{Höl}}^{\mu}$, tends to the one provided by Theorem 6.12 (the pink region). As a consequence,

$$D_H = D_{\text{Höl}}^1 = D_{\text{Höl}}^{\mu_j} \subset D_{\text{Höl}}^{\mu_{j-1}} \subset \cdots \subset D_{\text{Höl}}^{\mu_0} \qquad \text{for} \qquad \mu_0 < \cdots < \mu_{j-1} < \mu_j = 1.$$

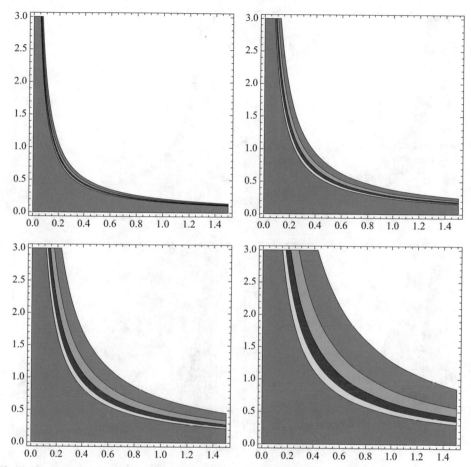

Fig. 6.12 Domains of parameters of Newton's method associated with Theorem 6.12 (the pink region) and Corollary 6.11 for $p = \frac{1}{10}, \frac{1}{5}, \frac{2}{5}, \frac{4}{5}$

Third, Fig. 6.13 shows that there exist values of $\mu \in (0, 1]$ such that

$$D_{\text{Höl}}^{\mu_1} \subset D_{\text{Höl}}^{\mu_2} \subset D_K \subset D_{\text{Höl}}^{\mu_3} \subset D_{\text{Höl}}^{\mu_4},$$

where $\mu_1, \mu_2, \mu_3, \mu_4 \in (0, 1]$, and we can guess that there exist values of $\mu \in (0, 1]$ for which $D_{\text{Höl}}^{\mu}$ is larger than D_K. In addition, $D_K \subset D_{\text{Höl}}^{\mu}$, for all $p \in (0.2856\ldots, 1]$ and $\mu \in (0, 1]$, and there exists $\epsilon \in (0, 1]$ such that $D_K \subset D_{\text{Höl}}^{\mu}$, for all $p \in [0, 0.2856\ldots]$ and $\mu < \epsilon$, and $D_{\text{Höl}}^{\mu} \subset D_K$, for all $p \in [0, 0.2856\ldots]$ and $\mu \in (\epsilon, 1]$. Notice that we can guess that the smaller the quantity $\mu = K_0/K \in (0, 1]$, the larger the domain of parameters: orange for $\mu = 0.1$, green for $\mu = 0.2$, red for $\mu = 0.4$ and yellow for $\mu = 0.8$.

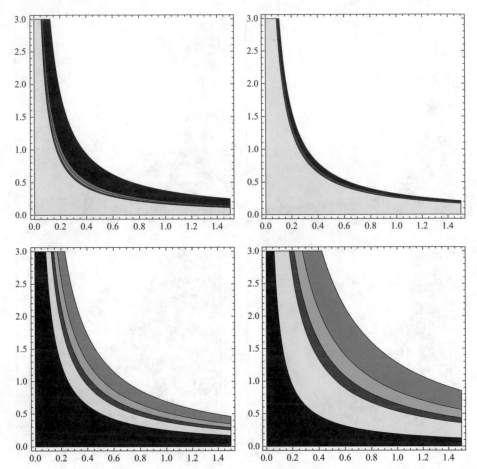

Fig. 6.13 Domains of parameters of Newton's method associated with Theorem 6.13 (the black region) and Corollary 6.11 for $p = \frac{1}{10}, \frac{1}{5}, \frac{2}{5}, \frac{4}{5}$

On the other hand, in Figs. 6.14, 6.15, and 6.16, we see how the domains of parameters associated with Corollary 6.11 and Theorems 6.12 and 6.13 compare as μ varies. For this, we place the values of μ on the x-axis and the values of $\theta_H = K\beta\eta^p$ in the y-axis. In Fig. 6.14, the cyan region is the domain associated with Theorem 6.13 and the brown region is the domain associated with Theorem 6.12. As we can see, sometimes one is larger and other times the other is larger, as it is proved in [48]. In Fig. 6.15, the magenta region represents the domain associated with Corollary 6.11. As we can see, the domain associated with Corollary 6.11 is always larger, for all $\mu \in (0, 1]$, than that associated with Theorem 6.12. Finally, in Fig. 6.16, we see that sometimes one is larger and other times the other is larger, but we can always find some values of $\mu \in (0, 1]$ such that the corresponding domain of parameters associated with Corollary 6.11 is greater than that associated with Theorem 6.13.

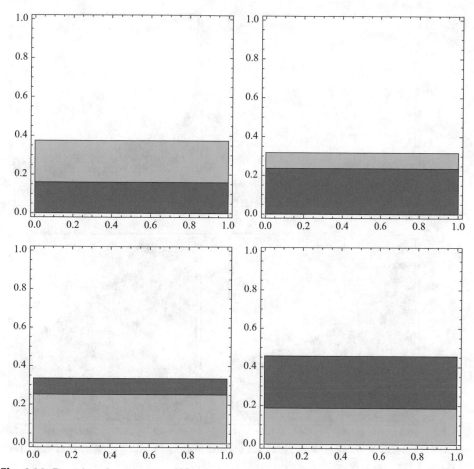

Fig. 6.14 Domains of parameters of Newton's method associated with Theorems 6.12 and 6.13 for $p = \frac{1}{10}, \frac{1}{5}, \frac{2}{5}, \frac{4}{5}$

In addition, in what follows we confirm the graphical conclusions. First, we prove that $D_H \subset D_{\text{Höl}}^{\mu}$, for all $p \in (0, 1]$ and $\mu \in (0, 1]$. For this, note that if $(K, \beta\eta^p) \in D_H$, then $\theta_H \leq \xi(p) \leq \frac{1}{2} < \frac{1}{2\mu} < \frac{1+p}{\mu(2+p)}$, since $p \in (0, 1]$ and $\mu \in (0, 1]$. Moreover, $\theta_H \leq \zeta(p) \leq \xi(p)\frac{(1+p)^p(1-\mu\theta_H)^{1+p}}{\mu^p\theta_H^p}$, since $\phi(\mu\theta_H; p) \geq 0$ when $\theta_H \leq \xi(p)$ and $\mu\theta_H \leq \xi(p)$. Therefore, $(K, \beta\eta^p) \in D_{\text{Höl}}^{\mu}$ for all $p \in (0, 1]$ and $\mu \in (0, 1]$.

Second, as $D_K \subset D_H$, for all $p \in (0.2856\ldots, 1]$ (see [48]), then $D_K \subset D_{\text{Höl}}^{\mu}$, for all $p \in (0.2856\ldots, 1]$ and $\mu \in (0, 1]$.

Third, we see that there exists $\epsilon \in [0, 1]$, for $p \in [0, 0.2856\ldots]$, such that $D_K \subset D_{\text{Höl}}^{\mu}$, for all $\mu \in (0, \epsilon]$. For this, note that if $(K, \beta\eta^p) \in D_K$, then $\theta_H \leq \frac{1}{2+p}\left(\frac{p}{1+p}\right)^p \leq \frac{1+p}{\mu(2+p)}$, since $\mu p^p \leq (1 + p)^{1+p}$, for all $\mu \in (0, 1]$. Moreover, as the condition $\theta_H \leq$

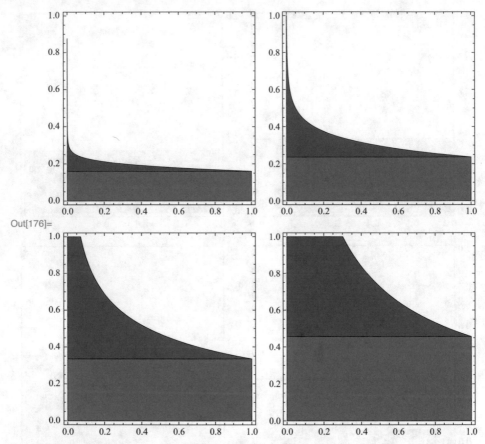

Out[176]=

Fig. 6.15 Domains of parameters of Newton's method associated with Corollary 6.11 and Theorem 6.12 for $p = \frac{1}{10}, \frac{1}{5}, \frac{2}{5}, \frac{4}{5}$

$\xi(p)\frac{(1+p)^p(1-\mu\theta_H)^{1+p}}{\mu^p\theta_H^p}$ is equivalent to

$$\theta_H \le \frac{\mu^{\frac{1}{1+p}}\xi(p)^{\frac{1}{1+p}}(1+p)^{\frac{p}{1+p}}}{\mu\left(1+\mu^{\frac{1}{1+p}}\xi(p)^{\frac{1}{1+p}}(1+p)^{\frac{p}{1+p}}\right)},$$

we deduce that

$$\frac{1}{2+p}\left(\frac{p}{1+p}\right)^p \le \xi(p)\frac{(1+p)^p(1-\mu\theta_H)^{1+p}}{\mu^p\theta_H^p}$$

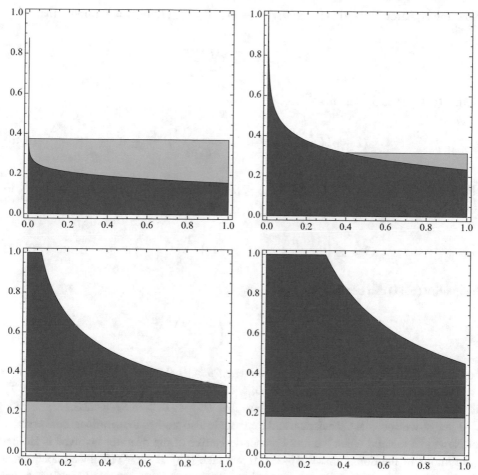

Fig. 6.16 Domains of parameters of Newton's method associated with Corollary 6.11 and Theorem 6.13 for $p = \frac{1}{10}, \frac{1}{5}, \frac{2}{5}, \frac{4}{5}$

is equivalent to

$$\frac{1}{2+p}\left(\frac{p}{1+p}\right)^p \leq \frac{(1+p)^{\frac{p}{1+p}}(\mu\xi(p))^{\frac{1}{1+p}}}{\mu\left(1+(1+p)^{\frac{p}{1+p}}(\mu\xi(p))^{\frac{1}{1+p}}\right)},$$

which, in turn, is equivalent to

$$(1+p)^{\frac{p}{1+p}}(\mu\xi(p))^{\frac{1}{1+p}}\left((2+p)(1+p)^p - \mu p^p\right) - \mu p^p \geq 0. \tag{6.45}$$

Now, the equation $(1 + p)^p (1 - \xi(p))^{1+p} - \xi(p)^p = \phi(\xi(p); p) = 0$ implies that

$$(1 + p)^{\frac{p}{1+p}} = \frac{\xi(p)^{\frac{p}{1+p}}}{1 - \xi(p)}$$

so that (6.45) is equivalent to

$$(2 + p)\left(\frac{1+p}{p}\right)^p - \mu - \frac{1 - \xi(p)}{\xi(p)}\mu^{\frac{p}{1+p}} \geq 0, \tag{6.46}$$

since $\frac{\xi(p)}{1-\xi(p)}\mu^{\frac{1}{1+p}}p^p \geq 0$. Moreover, as $\mu \leq \mu^{\frac{p}{1+p}}$, since $\mu \in (0, 1]$ and $\frac{p}{1+p} \in \left[0, \frac{1}{2}\right]$, to prove (6.46) it is sufficient that

$$(2 + p)\left(\frac{1+p}{p}\right)^p - \mu^{\frac{p}{1+p}}\left(1 + \frac{1 - \xi(p)}{\xi(p)}\right) \geq 0. \tag{6.47}$$

Inequality (6.47) can be rewritten as

$$\mu \leq \xi(p)^{\frac{1+p}{p}}(2 + p)^{\frac{1+p}{p}}\left(\frac{1+p}{p}\right)^{1+p} \equiv \epsilon,$$

so that the existence of $\epsilon \in [0, 1]$ is proved when $p \in [0, 0.2856\ldots]$.

After that, the proof of the inclusion $D_{\text{Höl}}^{\mu_2} \subset D_{\text{Höl}}^{\mu_1}$ for $\mu_1 < \mu_2$, with $\mu_1, \mu_2 \in (0, 1]$, is completely analogous to that used to show that $D_{\text{GC}}^{\mu_2} \subset D_{\text{GC}}^{\mu_1}$ in Sect. 6.2.2.1.

Finally, we note that Cianciaruso and De Pascale also studied the semilocal convergence of Newton's method when the first Fréchet derivative of the operator involved is Hölder continuous in the domain, see [21–23]. If we take [23] as their reference paper, since it is the most current and proposes a further improvement, we see that Cianciaruso and De Pascale establish the semilocal convergence of Newton's method from estimates of majorizing sequences and under the condition

$$\theta_{\text{H}} \leq \left(1 - \frac{1}{c}\right)^p \frac{1 + p}{\left((1+p)^{\frac{1}{1-p}} + (c(c - 1)^p)^{\frac{1}{1-p}}\right)^{1-p}} \quad \text{with} \quad c = \frac{1}{2}\sqrt{1 + 4(1 + p)^p p^{1-p}}.$$

If we now compare this condition with the conditions (6.44) required in Corollary 6.11, we see in the second column of Table 6.6 the values of $\mu \in (0, 1]$ to which Corollary 6.11 improves the estimates given by Cianciaruso and De Pascale in [23]. As we can see, for $p \geq 0.6$, Corollary 6.11 always improves the Cianciaruso and De Pascale estimates.

Table 6.6 Values of
$\mu \in (0, 1]$ to which
Corollary 6.11 improves the
estimates given by Cianciaruso
and De Pascale in [23]

p	μ
0.2	$[0, 0.0054\ldots]$
0.4	$[0, 0.3109\ldots]$
0.6	$[0, 1]$
0.8	$[0, 1]$
1	$[0, 1]$

Example 6.14 Consider the law $\Psi(x(t)) = 2 + x(t)^{\frac{5}{3}}$ for heat generation in problem
(6.18)–(6.19) and analyze the Hölder case studied before. So, the vector $v_{\mathbf{x}}$ of (6.21) is
given by

$$v_{\mathbf{x}} = (v_1, v_2, \ldots, v_8)^t, \qquad v_i = 2 + x_i^{\frac{5}{3}}, \qquad i = 1, 2, \ldots, 8, \qquad (6.48)$$

if $\mathfrak{m} = 8$ is chosen. In this case, we observe that a solution $\mathbf{x}^*$ of (6.21)–(6.48) satisfies

$$\|\mathbf{x}^*\| \leq \kappa^2 \|A^{-1}\| \|v_{\mathbf{x}^*}\| \qquad \Longrightarrow \qquad \|\mathbf{x}^*\| - \kappa^2 \|A^{-1}\| \left(2 + \|\mathbf{x}^*\|^{\frac{5}{3}}\right) \leq 0,$$

where $\|A^{-1}\| = 10$ and $\kappa = \frac{1}{9}$, so that $\|\mathbf{x}^*\| \in [0, \rho_1] \cup [\rho_2, +\infty)$, where $\rho_1 = 0.2599\ldots$ and $\rho_2 = 22.6775\ldots$, are the two positive real roots of the scalar equation
$81t - 10\left(2 + t^{\frac{5}{3}}\right) = 0$. In this case, we can consider

$$\mathbb{F}: \Lambda \subset \mathbb{R}^8 \to \mathbb{R}^8 \qquad \text{with} \qquad \Lambda = \{\mathbf{x} \in \mathbb{R}^8 : \|\mathbf{x}\| < 2\},$$

since $\rho_1 < 2 < \rho_2$.

Furthermore,

$$\mathbb{F}'(\mathbf{x}) = A + \kappa^2 \text{diag}(v'_{\mathbf{x}}), \quad v'_{\mathbf{x}} = (v'_1, v'_2, \ldots, v'_8)^T, \quad v'_i = \frac{5}{3}x_i^{\frac{2}{3}}, \quad i = 1, 2, \ldots, 8,$$

and

$$\mathbb{F}'(\mathbf{x}) - \mathbb{F}'(\mathbf{y}) = \frac{5}{3}\kappa^2 \text{diag}(\mathbf{z}),$$

where $\mathbf{y} = (y_1, y_2, \ldots, y_8)^T$ and $\mathbf{z} = \left(x_1^{\frac{2}{3}} - y_1^{\frac{2}{3}}, x_2^{\frac{2}{3}} - y_2^{\frac{2}{3}}, \ldots, x_8^{\frac{2}{3}} - y_8^{\frac{2}{3}}\right)^T$. Therefore,

$$\|\mathbb{F}'(\mathbf{x}) - \mathbb{F}'(\mathbf{y})\| \leq \frac{5}{3}\kappa^2 \left\|x^{\frac{1}{3}} + y^{\frac{1}{3}}\right\| \left\|x^{\frac{1}{3}} - y^{\frac{1}{3}}\right\| \leq \frac{10\sqrt[3]{2}}{3}\kappa^2 \|\mathbf{x} - \mathbf{y}\|^{\frac{1}{3}},$$

$$\|\mathbb{F}'(\mathbf{x}) - \mathbb{F}'(\mathbf{x_0})\| \leq \frac{5}{3}\kappa^2 \left(\|\mathbf{x}\|^{\frac{1}{3}} + \|\mathbf{x_0}\|^{\frac{1}{3}}\right) \|\mathbf{x} - \mathbf{x_0}\|^{\frac{1}{3}} \leq \frac{5}{3}\kappa^2 \left(\sqrt[3]{2} + \|\mathbf{x_0}\|^{\frac{1}{3}}\right) \|\mathbf{x} - \mathbf{x_0}\|^{\frac{1}{3}}.$$

Fig. 6.17 Domains of parameters associated with Theorem 6.12 (the blue region) and Corollary 6.11 when $\mu = \frac{1}{2}$ (pink and blue regions) and Newton's method is applied to solve the equation $\mathbb{F}(\mathbf{x}) = 0$ with $\mathbb{F}$ given in (6.21) and $v_{\mathbf{x}}$ in (6.48)

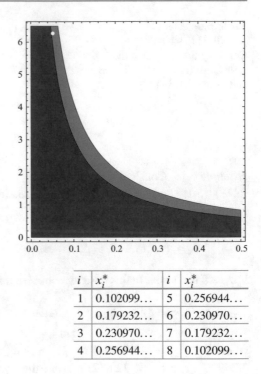

Table 6.7 Approximation of the solution $\mathbf{x}^*$ of (6.21)–(6.22)

i	x_i^*	i	x_i^*
1	0.102099...	5	0.256944...
2	0.179232...	6	0.230970...
3	0.230970...	7	0.179232...
4	0.256944...	8	0.102099...

Thus, $K = \frac{10\sqrt[3]{2}}{243} = 0.0518\ldots$, $K_0 = \frac{5}{243}\left(\sqrt[3]{2} + \|\mathbf{x}_0\|^{\frac{1}{3}}\right)$ and $p = \frac{1}{3}$.

If we choose the starting point $\mathbf{x}_0 = (0, 0, \ldots, 0)^T$, we obtain $\beta = 10$ and $\eta = 0.2469\ldots$, so that the condition of Theorem 6.12, $\theta_H \leq \xi(p)$, is not satisfied, since $\theta_H = 0.3252\ldots > \xi(p) = \xi(1/3) = 0.3071\ldots$ Therefore, according to Theorem 6.12, we cannot apply Newton's method for approximating a solution of (6.21)–(6.48).

However, we can guarantee the convergence of Newton's method from Corollary 6.11, since $K_0 = 0.0259\ldots$, $\mu = \frac{1}{2}$, $\theta_H = 0.3252\ldots$ and the conditions $\theta_H \leq \frac{1+p}{\mu(2+p)} = \frac{8}{7}$ and $\theta_H \leq \xi(p)(1+p)^p \frac{(1-\mu\theta_H)^{1+p}}{\mu^p\theta_H^p} = 0.4888\ldots$ of the corollary are satisfied. In Fig. 6.17, we see graphically that $\mathbf{x}_0$ is a good starting point for applying Newton's method, since its corresponding pair $(K, \beta\eta^p) = (0.0518\ldots, 6.2735\ldots)$, which is colored white, is in the domain of parameters associated with Corollary 6.11; see the white point. So, we can now apply Newton's method for approximating a solution of (6.21)–(6.48). We then obtain the approximation given by the vector $\mathbf{x}^* = (x_1^*, x_2^*, \ldots, x_8^*)^T$ and shown in Table 6.7, which is reached after four iterations. Observe that $\|\mathbf{x}^*\| = 0.2569\ldots \leq 2$, so that $\mathbf{x}^* \in \Lambda$. Table 6.8 displays the errors $\|\mathbf{x}_n - \mathbf{x}^*\|$ and the sequence $\{\|\mathbb{F}(\mathbf{x}_n)\|\}$. Notice that the vector shown in Table 6.7 is a good approximation of the solution of the system (6.21)–(6.22).

Table 6.8 Absolute errors
obtained by Newton's method
and the sequence $\{\|\mathbb{F}(\mathbf{x}_n)\|\}$

n	$\|\mathbf{x}_n - \mathbf{x}^*\|$	$\|\mathbb{F}(\mathbf{x}_n)\|$
0	$2.5694\ldots \times 10^{-1}$	$2.4691\ldots \times 10^{-2}$
1	$1.0030\ldots \times 10^{-2}$	$1.1997\ldots \times 10^{-3}$
2	$8.6683\ldots \times 10^{-6}$	$1.0932\ldots \times 10^{-6}$
3	$6.3598\ldots \times 10^{-12}$	$8.1065\ldots \times 10^{-13}$

Operators with Center ω-Lipschitz Continuous First Derivative

7

There is a rich literature concerning Newton-Kantorovich type results on the semilocal convergence of Newton's method. Many authors have studied the semilocal convergence of the method under different modifications of condition (A2):

$$\|F'(x) - F'(y)\| \le L\|x - y\|, \quad x, y \in \Omega. \tag{7.1}$$

Instead of assuming (A2), i.e., that the first Fréchet derivative F' is Lipschitz continuous in Ω, one can assume, as in Chap. 3 that F' is Hölder continuous in Ω or, more generally, as in Chap. 5, that

$$\|F'(x) - F'(y)\| \le \omega(\|x - y\|), \quad x, y \in \Omega, \tag{7.2}$$

where $\omega : [0, +\infty) \to \mathbb{R}$ is a continuous and nondecreasing function such that $\omega(0) \ge 0$. The aim of this chapter is to prove the semilocal convergence of Newton's method under milder differentiability conditions than the previous ones. For this purpose, we use a center ω-Lipschitz condition for F' instead of condition (7.2). So, we fix one of the two points of condition (7.2), so that this point is the starting point x_0 of Newton's method and the condition is then satisfied for any $x \in \Omega$ and x_0 instead of any two points $x, y \in \Omega$.

There are situations where F' is only ω-Lipschitz continuous in some points of the domain, but not in all the domain. This justifies the study of center conditions for F', as we can see in the following application, where we propose to relax condition (7.1) by condition (6.2),

$$\|F'(x) - F'(x_0)\| \le L_0\|x - x_0\|, \quad x \in \Omega. \tag{7.3}$$

© The Editor(s) (if applicable) and The Author(s), under exclusive licence to Springer Nature Switzerland AG 2020
J. A. Ezquerro Fernández, M. Á. Hernández Verón, *Mild Differentiability Conditions for Newton's Method in Banach Spaces*, Frontiers in Mathematics, https://doi.org/10.1007/978-3-030-48702-7_7

Obviously, condition (7.1) implies (7.3), while the converse is not true, so that there are operators that satisfy condition (7.3), but not condition (7.1), as we can see in the following application.

Nonlinear elliptic equations are usually difficult to treat numerically. However, mildly nonlinear elliptic equations of the form:

$$u_{xx} + u_{yy} = \Psi(u), \tag{7.4}$$

where Ψ is a continuous nonlinear scalar function, are an exception. This type of equations are of interest in gas dynamics [42]. An associated Dirichlet problem can be formulated as follows. We consider the Eq. (7.4) in the rectangular domain $\mathcal{D} = \{(x, y) \in \mathbb{R}^2 : a \leq x \leq b, c \leq y \leq d\}$ with the Dirichlet boundary conditions

$$u(x, c) = f_1(x), \; u(x, d) = f_2(x), \; a \leq x \leq b,$$
$$u(a, y) = g_1(y), \; u(b, y) = g_2(y), \; c \leq y \leq d. \tag{7.5}$$

Next, we formulate a finite difference approximation for the elliptic boundary value problem (7.4)–(7.5). The method of finite differences applied to this problem yields a finite system of equations. For general use, iterative techniques often represent the best approach to the solution of such finite systems of equations.

Specifically, we use central difference approximations for (7.4), so that problem (7.4)–(7.5) is reduced to the problem of finding a real zero of a function $\mathbb{F} : \Lambda \subseteq \mathbb{R}^m \to \mathbb{R}^m$, namely a real solution $\mathbf{x}^*$ of a nonlinear system $\mathbb{F}(\mathbf{x}) = 0$ with m equations and m unknowns. We then approximate $\mathbf{x}^*$ bu Newton's method, since this method is one of the most efficient.

For problem (7.4)–(7.5) in $\mathbb{R}^2$, Eq. (7.4) can be approximated using a central difference for the spatial derivatives. Consider a grid with step size $\mathfrak{s} = \frac{b-a}{N+1}$, $N \in \mathbb{N}$, in x and $\mathfrak{t} = \frac{d-c}{M+1}$, $M \in \mathbb{N}$, in y defined on the domain $\mathcal{D}$, so that $\mathcal{D}$ is partitioned into a grid consisting of $(N + 1) \times (M + 1)$ rectangles with sides $\mathfrak{s}$ and $\mathfrak{t}$. The mesh points (x_i, y_j) are given by

$$x_i = a + i\mathfrak{s}, \quad y_i = c + j\mathfrak{t}, \quad i = 0, 1, \ldots, N+1, \quad j = 0, 1, \ldots, M+1.$$

Let us use the following finite difference expressions to approximate the partial differentials:

$$u_{xx}(x_i, y_j) = \frac{u(x_{i-1}, y_j) - 2u(x_i, y_j) + u(x_{i+1}, y_j)}{\mathfrak{s}^2} + \mathcal{O}(\mathfrak{s}^2),$$

$$u_{yy}(x_i, y_j) = \frac{u(x_i, y_{j-1}) - 2u(x_i, y_j) + u(x_i, y_{j+1})}{\mathfrak{t}^2} + \mathcal{O}(\mathfrak{t}^2).$$

Then, Eq. (7.4) is approximated at each interior grid point (x_i, y_j) by the difference equation

$$\frac{u(x_{i+1}, y_j) - 2u(x_i, y_j) + u(x_{i-1}, y_j)}{s^2} + \frac{u(x_i, y_{j+1}) - 2u(x_i, y_j) + u(x_i, y_{j-1})}{t^2} = \Psi(u(x_i, y_j)),$$

for $i = 1, 2, \ldots, N$ and $j = 1, 2, \ldots, M$. The boundary conditions are

$$u(x_i, y_0) = f_1(x_i), \quad u(x_i, y_M) = f_2(x_i), \quad i = 1, 2, \ldots, N,$$

$$u(x_0, y_j) = g_1(y_j), \quad u(x_N, y_j) = g_2(y_j), \quad j = 0, 1, \ldots, M+1.$$

Denoting the approximate value of $u(x_i, y_j)$ by $u_{i,j}$, we obtain the difference equation

$$2\left[\left(\frac{s}{t}\right)^2 + 1\right] u_{i,j} - (u_{i-1,j} + u_{i+1,j}) - \left(\frac{s}{t}\right)^2 (u_{i,j-1} + u_{i,j+1}) = -s^2 \Psi(u_{i,j}),$$

for $i = 1, 2, \ldots, N$ and $j = 1, 2, \ldots, M$, with

$$u_{i,0} = f_1(x_i), \quad u_{i,M+1} = f_2(x_i), \quad i = 1, 2, \ldots, N,$$

$$u_{0,j} = g_1(y_j), \quad u_{N+1,j} = g_2(y_j), \quad j = 0, 1, \ldots, M+1.$$

Equation (7.4) with the boundary conditions (7.5) yields an $NM \times NM$ nonlinear system of equations. To set up the nonlinear system, the $NM = m$ interior grid points are labeled row-by-row from x_1 to x_m starting from the left-bottom corner point. The resulting system reads

$$A\mathbf{x} + s^2 q(\mathbf{x}) = \mathbf{v},$$

where

$$A = \begin{pmatrix} B & C & 0 & \cdots\cdots & 0 \\ C & B & C & \ddots & \vdots \\ 0 & C & B & C & \ddots & \vdots \\ \vdots & \ddots & \ddots & \ddots & \ddots & 0 \\ \vdots & & \ddots & \ddots & \ddots & C \\ 0 & \cdots\cdots & & 0 & C & B \end{pmatrix}_{M \times M},$$

$$B = \begin{pmatrix} 2(\lambda+1) & -1 & 0 & \cdots & \cdots & 0 \\ -1 & 2(\lambda+1) & -1 & \ddots & & \vdots \\ 0 & -1 & 2(\lambda+1) & -1 & \ddots & \vdots \\ \vdots & \ddots & \ddots & \ddots & \ddots & 0 \\ \vdots & & \ddots & \ddots & \ddots & -1 \\ 0 & \cdots & \cdots & 0 & -1 & 2(\lambda+1) \end{pmatrix}_{N \times N},$$

$C = -\lambda I$, $\lambda = \left(\frac{s}{t}\right)^2$, I is the $N \times N$ identity matrix, $\mathbf{x}$ is the vector $(x_1, x_2, \ldots, x_{\mathfrak{m}})^t$, $q(\mathbf{x}) = (\Psi(x_1), \Psi(x_2), \ldots, \Psi(x_{\mathfrak{m}}))^t$ and $\mathbf{v}$ is a vector formed from the boundary conditions. Systems of this type are so-called mildly nonlinear.

If we write our system in the form $\mathbb{F}(\mathbf{x}) = 0$, where

$$\mathbb{F}(\mathbf{x}) = A\mathbf{x} + s^2 q(\mathbf{x}) - \mathbf{v}, \qquad \mathbb{F} : \mathbb{R}^{\mathfrak{m}} \longrightarrow \mathbb{R}^{\mathfrak{m}}, \qquad \mathfrak{m} = NM, \qquad (7.6)$$

then $\mathbb{F}'(\mathbf{x})$ is the linear operator given by the matrix

$$\mathbb{F}'(\mathbf{x}) = A + s^2 Q(\mathbf{x}), \qquad Q(\mathbf{x}) = \mathrm{diag}\left\{\Psi'(x_1), \Psi'(x_2), \ldots, \Psi'(x_{\mathfrak{m}})\right\}.$$

We illustrate this discussion with the equation

$$u_{xx} + u_{yy} = u^{\frac{3}{2}}. \qquad (7.7)$$

The study of this equation for a bounded region is of interest in itself; the equation has no entire solutions (i.e., solutions without singularities stet a finite distance).

In this case, $\mathbb{F}'(\mathbf{x})$ is the linear operator given by the matrix

$$A + \frac{3}{2} s^2 \, \mathrm{diag}\left\{x_1^{\frac{1}{2}}, x_2^{\frac{1}{2}}, \ldots, x_{\mathfrak{m}}^{\frac{1}{2}}\right\},$$

so that

$$\mathbb{F}'(\mathbf{x}) - \mathbb{F}'(\mathbf{y}) = \frac{3}{2} s^2 \, \mathrm{diag}\left\{x_1^{\frac{1}{2}} - y_1^{\frac{1}{2}}, x_2^{\frac{1}{2}} - y_2^{\frac{1}{2}}, \ldots, x_{\mathfrak{m}}^{\frac{1}{2}} - y_{\mathfrak{m}}^{\frac{1}{2}}\right\},$$

where $\mathbf{y} = (y_1, y_2, \ldots, y_{\mathfrak{m}})^T$.

Notice that $\mathbb{F}'$ is not Lipschitz continuous in all the domain $\mathbb{R}^{\mathfrak{m}}$. However, if we choose $\mathbf{x}_0 = \mathbf{1} = (1, 1, \ldots, 1)^T \in \mathbb{R}^{\mathfrak{m}}$, then

$$\|\mathbb{F}'(\mathbf{x}) - \mathbb{F}'(\mathbf{x}_0)\| = \|\mathbb{F}'(\mathbf{x}) - \mathbb{F}'(\mathbf{1})\| \leq \frac{3}{2} s^2 \|\mathbf{x} - \mathbf{1}\| = \frac{3}{2} s^2 \|\mathbf{x} - \mathbf{x}_0\|,$$

since $\left| x_i^{\frac{1}{2}} - 1 \right| = \frac{|x_i - 1|}{x_i^{\frac{1}{2}} + 1} \leq |x_i - 1|$, for all $i = 1, 2, \ldots, \mathfrak{m}$, so that $\mathbb{F}'$ is Lipschitz continuous at $\mathbf{x}_0$.

Thus, an important consequence of the above discussion is that the semilocal convergence results for Newton's method under center conditions for $\mathbb{F}'$ are interesting regardless of the general non-centered situation.

However, we observe that condition $L_0 \beta \eta \leq \zeta$ with $\zeta < \frac{1}{2}$ must be satisfied if condition (7.3) with $L_0 \leq L$ holds if one wants to guarantee the semilocal convergence of Newton's method under a center Lipschitz condition for the first Fréchet derivative $\mathbb{F}'$, as we can see later in Corollary 7.5. In addition, this condition could be more restrictive than the condition $L \beta \eta \leq \frac{1}{2}$ required in the Lipschitz case. On the other hand, since $L_0 \leq L$, we could verify condition $L_0 \beta \eta \leq \zeta$, but not $L \beta \eta \leq \frac{1}{2}$. In this case, new starting points for Newton's method can be found, as we can show below.

7.1 Convergence Analysis

In this section, we follow [46] and investigate whether it is possible to relax condition (7.3) to

$$\| F'(x) - F'(x_0) \| \leq \omega_0(\|x - x_0\|), \quad x \in \Omega, \tag{7.8}$$

where the function $\omega_0 : [0, +\infty) \to \mathbb{R}$ is continuous, nondecreasing and such that $\omega_0(0) \geq 0$.

First, we notice that (7.3) obviously implies (7.8). However, the reciprocal is not true. Next, we suppose the following conditions:

(P1) There exists the operator $\Gamma_0 = [F'(x_0)]^{-1} \in \mathcal{L}(Y, X)$, for some $x_0 \in \Omega$, with $\|\Gamma_0\| \leq \beta$ and $\|\Gamma_0 F(x_0)\| \leq \eta$.

(P2) $\| F'(x) - F'(x_0) \| \leq \omega_0(\|x - x_0\|)$, $x \in \Omega$, where $\omega_0 : [0, +\infty) \to \mathbb{R}$ is a nondecreasing continuous function such that $\omega_0(0) \geq 0$.

(P3) There exists a nondecreasing continuous function $h_0 : [0, 1] \to \mathbb{R}$, such that $\omega_0(tz) \leq h_0(t)\omega_0(z)$ for $t \in [0, 1]$ and $z \in [0, \infty)$.

(P4) The equation

$$g(t) = (1 - (3 - I_{h_0})\omega_0(t))\eta - (1 - 3\omega_0(t))t = 0,$$

where $I_{h_0} = \int_0^1 h_0(\tau) \, d\tau$, has at least one positive root. We denote the smallest positive root of this equation by R.

(P5) $\beta\omega_0(R) < \frac{1}{3}$ and $B(x_0, R) \subset \Omega$.

Notice that condition (P3) is not restrictive, since there always exists an h_0, such that $h_0(t) = 1$, because ω_0 is a nondecreasing function. It is used to sharpen the bounds that we obtain for particular expressions.

Next, we analyse the semilocal convergence of Newton's method under conditions (P1)-(P2)-(P3)-(P4)-(P5) and use the theoretical power of the method to draw conclusions about the existence and uniqueness of solution. We start with two technical lemmas that will be needed below.

Lemma 7.1 *If conditions the* (P1)-(P2)-(P3)-(P4)-(P5) *are satisfied and* $f(t) = \frac{1}{1-\beta\omega_0(t)}$, *then*

$$\left(1 + I_{h_0} \sum_{j=1}^{k} 2^{j-1}(\beta\omega_0(R)f(R))^j\right)\eta < \left(1 + \frac{I_{h_0}\beta\omega_0(R)f(R)}{1 - 2\beta\omega_0(R)f(R)}\right)\eta$$

$$= \frac{(1 - (3 - I_{h_0})\beta\omega_0(R))\eta}{1 - 3\beta\omega_0(R)} = R.$$

Proof Observe that the first expression includes the sum of a geometric series with ratio $2\beta\omega_0(R)f(R)$. If $\beta\omega_0(R) < \frac{1}{3}$, this ratio is < 1 and the inequality follows immediately. Next, the equality follows from condition (P4). ∎

Lemma 7.2 *Suppose the conditions* (P1)-(P2)-(P3)-(P4)-(P5). *With the notations of Lemma* 7.1, *we can assert that:*

(I_n) *There exists the operator* $\Gamma_n = [F'(x_n)]^{-1}$ *and* $\|\Gamma_n F'(x_0)\| \le f(R)$, $n \ge 1$.

(II_n) $\|\Gamma_0 F(x_n)\| \le 2\beta\omega_0(R)\|x_n - x_{n-1}\|$, $n \ge 2$.

(III_n) $\|x_{n+1} - x_n\| \le 2\beta\omega_0(R)f(R)\|x_n - x_{n-1}\|$, $n \ge 2$.

(IV_n) $\|x_{n+1} - x_0\| \le \left(1 + I_{h_0} \sum_{i=0}^{n-1} 2^i (\beta\omega_0(R)f(R))^{i+1}\right)\eta < R$, $n \ge 1$.

Proof If $x \in B(x_0, R)$, then

$$\|I - \Gamma_0 F'(x)\| = \|\Gamma_0(F'(x) - F'(x_0))\| \le \beta\omega_0(\|x - x_0\|) \le \beta\omega_0(R) < 1.$$

As a consequence, the operator $[F'(x)]^{-1}$ exists and

$$\|[F'(x)]^{-1}F'(x_0)\| \le \frac{1}{1 - \beta\omega_0(R)} = f(R).$$

Next, from (P5) and $\|x_1 - x_0\| \le \eta < R$, it follows that $x_1 \in B(x_0, R) \subset \Omega$ and $\|\Gamma_1 F'(x_0)\| \le f(R)$. Then a Taylor series expansion yields that, we have

$$
\begin{aligned}
\|\Gamma_0 F(x_1)\| &= \left\| \int_{x_0}^{x_1} \Gamma_0(F'(x) - F'(x_0))\, dx \right\| \\
&= \left\| \int_0^1 \Gamma_0(F'(x_0 + \tau(x_1 - x_0)) - F'(x_0))(x_1 - x_0)\, d\tau \right\| \\
&\le \beta \int_0^1 \omega_0(\tau \|x_1 - x_0\|)\, d\tau \|x_1 - x_0\| \\
&\le \beta \int_0^1 h_0(\tau)\, d\tau\, \omega_0(R) \|x_1 - x_0\| \\
&= I_{h_0} \beta \omega_0(R) \|x_1 - x_0\|
\end{aligned}
$$

and

$$
\|x_2 - x_1\| \le \|\Gamma_1 F'(x_0)\| \|\Gamma_0 F(x_1)\| \le I_{h_0} \beta \omega_0(R) f(R) \|x_1 - x_0\|. \tag{7.9}
$$

Then, in view of Lemma 7.1,

$$
\begin{aligned}
\|x_2 - x_0\| &\le \|x_2 - x_1\| + \|x_1 - x_0\| \\
&\le \left(1 + I_{h_0} \beta \omega_0(R) f(R)\right) \|x_1 - x_0\| \tag{7.10} \\
&\le \left(1 + I_{h_0} \beta \omega_0(R) f(R)\right) \eta \\
&< R
\end{aligned}
$$

and $x_2 \in B(x_0, R) \subset \Omega$.

Next, the operator Γ_2 exists and

$$
\|\Gamma_2 F'(x_0)\| \le f(R).
$$

Since $x_1 + \tau(x_2 - x_1) \in B(x_0, R)$ with $\tau \in [0, 1]$, we see that

$$
\begin{aligned}
\|\Gamma_0 F(x_2)\| &= \left\| \int_{x_1}^{x_2} \Gamma_0(F'(x) - F'(x_1))\, dx \right\| \\
&= \left\| \int_0^1 \Gamma_0(F'(x_1 + \tau(x_2 - x_1)) \pm F'(x_0) - F'(x_1))(x_2 - x_1)\, d\tau \right\| \\
&\le 2\beta \omega_0(R) \|x_2 - x_1\|
\end{aligned}
$$

and

$$\|x_3 - x_2\| \leq \|\Gamma_2 F'(x_0)\| \|\Gamma_0 F(x_2)\| \leq 2\beta\omega_0(R) f(R) \|x_2 - x_1\|.$$

Lemma 7.1 and the inequalities (7.9) and (7.10) imply that

$$
\begin{aligned}
\|x_3 - x_0\| &\leq \|x_3 - x_2\| + \|x_2 - x_0\| \\
&\leq 2\beta\omega_0(R) f(R) \|x_2 - x_1\| + \left(1 + I_{h_0}\beta\omega_0(R) f(R)\right) \eta \\
&\leq \left(1 + I_{h_0}\beta\omega_0(R) f(R) + 2I_{h_0}(\beta\omega_0(R) f(R))^2\right) \eta \\
&< R
\end{aligned}
$$

and $x_3 \in B(x_0, R)$.

If we now suppose that items (I_n)-(II_n)-(III_n)-(IV_n) hold for $n = 1, 2, \ldots, i - 1$, we see that they also hold for $n = 1$. Since $x_i \in B(x_0, R)$, the operator Γ_i exists and

$$\|\Gamma_i F'(x_0)\| \leq f(R).$$

Moreover,

$$\|\Gamma_0 F(x_i)\| \leq 2\beta\omega_0(R) \|x_i - x_{i-1}\| \tag{7.11}$$

and

$$
\begin{aligned}
\|x_i - x_{i-1}\| &\leq \|\Gamma_{i-1} F'(x_0)\| \|\Gamma_0 F(x_{i-1})\| \tag{7.12} \\
&\leq 2\beta\omega_0(R) f(R) \|x_{i-1} - x_{i-2}\| \\
&\leq \cdots \leq (2\beta\omega_0(R) f(R))^{i-2} \|x_2 - x_1\| \\
&\leq (2\beta\omega_0(R) f(R))^{i-2} I_{h_0}\beta\omega_0(R) f(R)\eta \\
&= 2^{i-2} I_{h_0} (\beta\omega_0(R) f(R))^{i-1} \eta.
\end{aligned}
$$

Hence, (7.11) and the fact that $\|x_i - x_0\| < R$ imply that

$$\|x_{i+1} - x_i\| \leq \|\Gamma_i F'(x_0)\| \|\Gamma_0 F(x_i)\| \leq 2\beta\omega_0(R) f(R) \|x_i - x_{i-1}\|.$$

Using Lemma 7.1, we conclude that

$$\|x_{i+1} - x_0\| \leq \|x_{i+1} - x_i\| + \|x_i - x_0\|$$

$$\leq \left(1 + I_{h_0} \sum_{j=0}^{i-2} 2^j \left(\beta\omega_0(R)f(R)\right)^{j+1}\right) \eta + 2\beta\omega_0(R)f(R)\|x_i - x_{i-1}\|$$

$$\leq \left(1 + I_{h_0} \sum_{j=0}^{i-1} 2^j \left(\beta\omega_0(R)f(R)\right)^{j+1}\right) \eta$$

$$< R,$$

which completes the induction step.　∎

The semilocal convergence of Newton's method under the conditions (P1)-(P2)-(P3)-(P4)-(P5) follows now from the preceding discussion.

Theorem 7.3 *Let* $F : \Omega \subseteq X \rightarrow Y$ *be a continuously Fréchet differentiable operator defined on a nonempty open convex domain* Ω *of a Banach space* X *with values in a Banach space* Y. *If conditions the (P1)-(P2)-(P3)-(P4)-(P5) hold, then the Newton sequence* $\{x_n\}$ *starting at* x_0 *converges to a solution* x^* *of equation* $F(x) = 0$. *Moreover,* $x_n, x^* \in \overline{B(x_0, R)}$ *and* x^* *is unique in* $\overline{B(x_0, R)}$.

Proof By Lemma 7.2, (7.11) and (7.12), we have

$$\|x_{n+m} - x_n\| \leq \sum_{j=n}^{n+m-1} \|x_{j+1} - x_j\|$$

$$\leq \sum_{j=n}^{n+m-1} \|\Gamma_j F'(x_0)\| \|\Gamma_0 F(x_j)\|$$

$$\leq f(R) \sum_{j=n}^{n+m-1} \|\Gamma_0 F(x_j)\|$$

$$\leq 2\beta\omega_0(R)f(R) \sum_{j=n}^{n+m-1} (2\beta\omega_0(R)f(R))^{j-2}\|x_2 - x_1\|$$

$$\leq \left(\sum_{j=n}^{n+m-1} 2^{j-1}(\beta\omega_0(R)f(R))^j\right) I_{h_0}\,\eta.$$

Hence, $\{x_n\}$ is a Cauchy sequence and therefore convergent, since $2\beta\omega_0(R)f(R) < 1$. Let $\lim x_n = x^*$. Letting $i \to \infty$ in (7.11) and using the continuity of F, we conclude that $F(x^*) = 0$.

Finally, we prove the uniqueness of the solution x^*. For this, we suppose that y^* is another solution of $F(x) = 0$ in $\overline{B(x_0, R)}$. Then, the approximation

$$0 = F(y^*) - F(x^*) = \int_{x^*}^{y^*} F'(x)dx = \int_0^1 F'(x^* + \tau(y^* - x^*))\, d\tau(y^* - x^*)$$

implies that $x^* = y^*$ because the operator $\int_0^1 F'(x^* + \tau(y^* - x^*))\, d\tau$ is invertible. The invertibility follows from the existence of J^{-1}, where $J = \Gamma_0 \int_0^1 F'(x^* + \tau(y^* - x^*))\, d\tau$, which in turn is a consequence of the inequality

$$\|I - J\| = \left\| \int_0^1 \Gamma_0[F'(y^* + \tau(x^* - y^*)) - F'(x_0)]\, d\tau \right\|$$

$$\leq \|\Gamma_0\| \int_0^1 \omega_0(\|y^* + \tau(x^* - y^*) - x_0\|)\, d\tau$$

$$\leq \beta\omega_0(R)$$

$$< 1$$

and the Banach lemma on invertible operators. The proof is complete. ∎

7.2   Application to an Elliptic Problem

In this section, we apply the previous result to a nonlinear elliptic equation of form (7.4).

Example 7.4 Consider the equation

$$u_{xx} + u_{yy} = u^3 + u^{\frac{5}{3}}. \tag{7.13}$$

Suppose that the equation is satisfied in the interior of the square $0 \leq x, y \leq 1$ in $\mathbb{R}^2$, and that $u(x, y) > 0$ is given and continuous on the boundary of the square and satisfies there some boundary conditions [70]

$$u(x, 0) = 2x^2 - x + 1, \, u(x, 1) = 2, \, 0 \leq x \leq 1,$$

$$u(0, y) = 2y^2 - y + 1, \, u(1, y) = 2, \, 0 \leq y \leq 1. \tag{7.14}$$

Next, we apply the method of finite differences, developed in the introduction of the chapter, to the nonlinear elliptic problem (7.13)–(7.14) with $N = M = 4$, so that $\mathfrak{s} = \frac{1}{5}$, $\lambda = 1$ and the number of equations is $\mathfrak{m} = NM = 16$. Moreover,

$$q(\mathbf{x}) = \left(x_1^3 + x_1^{\frac{5}{3}}, x_2^3 + x_2^{\frac{5}{3}}, \ldots, x_{16}^3 + x_{16}^{\frac{5}{3}} \right)^T \text{ and the vector } \mathbf{v} \text{ is}$$

$$\mathbf{v} = \left(\frac{44}{25}, \frac{23}{25}, \frac{28}{25}, \frac{87}{25}, \frac{23}{25}, 0, 0, 2, \frac{28}{25}, 0, 0, 2, \frac{87}{25}, 2, 2, 4 \right)^T t. \qquad (7.15)$$

In this case, we observe that a solution $\mathbf{x}^*$ of the system $\mathbb{F}(\mathbf{x}) = 0$, where $\mathbb{F}$ is defined in (7.6) with $\Psi(u) = u^3 + u^{\frac{5}{3}}$ necessarily satisfies

$$\|\mathbf{x}^*\| \leq \|A^{-1}\| \left(\|\mathbf{v}\| + \mathfrak{s}^2 \|q(\mathbf{x})\| \right),$$

since

$$\|\mathbf{x}^*\| - \frac{5}{3} \left(4 + \frac{1}{25} \left(\|\mathbf{x}^*\|^3 + \|\mathbf{x}^*\|^{5/3} \right) \right) \leq 0,$$

where $\|A^{-1}\| = \frac{5}{3}$ and $\|\mathbf{v}\| = 4$. Then, we can consider

$$\mathbb{F} : \Lambda \subset \mathbb{R}^{16} \longrightarrow \mathbb{R}^{16} \quad \text{with} \quad \Lambda = \left\{ \mathbf{x} \in \mathbb{R}^{16} : \|\mathbf{x}\| < \frac{2}{5} \right\}.$$

The linear operator $\mathbb{F}'(\mathbf{x})$ is given by the matrix

$$A + \mathfrak{s}^2 \operatorname{diag} \left\{ 3x_1^2 + \frac{5}{3} x_1^{1/3}, 3x_2^2 + \frac{5}{3} x_2^{1/3}, \ldots, 3x_{16}^2 + \frac{5}{3} x_{16}^{1/3} \right\},$$

and

$$\|\mathbb{F}'(\mathbf{x}) - \mathbb{F}'(\mathbf{x}_0)\| \leq \mathfrak{s}^2 \left(3 \left(\left(\frac{2}{5} \right) + \|\mathbf{x}_0\| \right) \|\mathbf{x} - \mathbf{x}_0\| + \frac{5}{3} \left(\left(\frac{2}{5} \right)^{\frac{1}{3}} + \|\mathbf{x}_0\|^{\frac{1}{3}} \right) \|\mathbf{x} - \mathbf{x}_0\|^{\frac{1}{3}} \right),$$

so that

$$\omega_0(t) = \mathfrak{s}^2 \left(3 \left(\left(\frac{2}{5} \right) + \|\mathbf{x}_0\| \right) t + \frac{5}{3} \left(\left(\frac{2}{5} \right)^{\frac{1}{3}} + \|\mathbf{x}_0\|^{\frac{1}{3}} \right) t^{\frac{1}{3}} \right).$$

Table 7.1 Approximation of the solution $\mathbf{x}^*$ of $\mathbb{F}(\mathbf{x}) = 0$ with $\mathbb{F}$ given in (7.6) and $\Psi(u) = u^3 + u^{\frac{5}{3}}$

i	x_i^*	i	x_i^*	i	x_i^*	i	x_i^*
1	0.93480628...	5	1.02382485...	9	1.20010680...	13	1.50114261...
2	1.02382485...	6	1.12491489...	10	1.27880992...	14	1.53849446...
3	1.20010680...	7	1.27880992...	11	1.39564152...	15	1.60170155...
4	1.50114261...	8	1.53849446...	12	1.60170155...	16	1.72473949...

Table 7.2 Absolute errors obtained by Newton's method and $\{\|\mathbb{F}(\mathbf{x}_n)\|\}$

n	$\|\mathbf{x}^* - \mathbf{x}_n\|$	$\|\mathbb{F}(\mathbf{x}_n)\|$
0	1.7247...	4
1	$3.0011... \times 10^{-1}$	$4.0307... \times 10^{-1}$
2	$1.3848... \times 10^{-2}$	$1.7295... \times 10^{-2}$
3	$2.4961... \times 10^{-5}$	$3.6008... \times 10^{-5}$
4	$7.5131... \times 10^{-11}$	$1.1674... \times 10^{-10}$

If we now choose $\mathbf{x}_0 = (0, 0, \ldots, 0)^T$, we obtain $\beta = 1.6666\ldots$, $\eta = 1.9224\ldots$ and the equation appearing in condition (P4) reduces to

$$g(t) = (0.144)t^2 + (0.1473\ldots)t^{\frac{4}{3}} - (1.2076\ldots)t - (0.2124\ldots)t^{\frac{1}{3}} + (1.9224\ldots) = 0,$$

which has two positive roots, the smallest of which, $R = 2.5601\ldots$, satisfies the condition appearing in (P5), $\beta\omega_0(R) = 0.3168\ldots < \frac{1}{3}$. Therefore, the conditions of Theorem 7.3 are satisfied, and so the system $\mathbb{F}(x) = 0$, where $\mathbb{F}$ is given in (7.6) with $\Psi(u) = u^3 + u^{\frac{5}{3}}$, has a unique solution $\mathbf{x}^*$ in the region $\{\mathbf{v} \in \mathbb{R}^{16} : \|\mathbf{v} - \mathbf{x}_0\| \leq 2.5601\ldots\}$ and the convergence of Newton's method to $\mathbf{x}^*$ is then guaranteed by Theorem 7.3.

We can then apply Newton's method to obtain an approximation of the solution $\mathbf{x}^*$ given in Table 7.1, which is reached after five iterations. Table 7.2 displays the errors $\|\mathbf{x}^* - \mathbf{x}_n\|$ and the sequence $\{\|\mathbb{F}(\mathbf{x}_n)\|\}$. Notice that the vector shown in Table 7.1 is a good approximation of a solution of nonlinear system under study.

Finally, we note that if we interpolate the points in Table 7.1 and take into account that the solution satisfies the boundary conditions in (7.14), we obtain the approximate numerical solution shown in Fig. 7.1.

7.3   Particular Case

The main particular case of operators with center ω-Lipschitz first Fréchet derivative that we already know is when F' is center Lipschitz continuous [43]. So, if $\omega_0(z) = L_0 z$ and $h_0(t) = t$ in (P3), F' is center Lipschitz continuous in Ω and the conditions (P1)-(P2)-

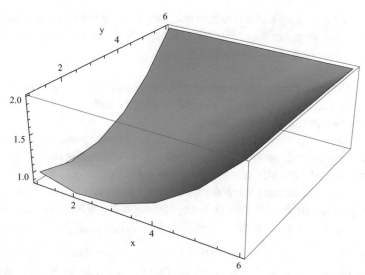

Fig. 7.1 Approximate solution of problem (7.13)–(7.14)

(P3)-(P4)-(P5) reduce to the following ones:

(Q1)   There exists the operator $\Gamma_0 = [F'(x_0)]^{-1} \in \mathcal{L}(Y, X)$, for some $x_0 \in \Omega$, with $\|\Gamma_0\| \leq \beta$ and $\|\Gamma_0 F(x_0)\| \leq \eta$;

(Q2) $\|F'(x) - F'(x_0)\| \leq L_0\|x - x_0\|$ for $x \in \Omega$;

(Q3) $\vartheta = L_0\beta\eta \leq \frac{14-4\sqrt{6}}{25} = 0.168081\ldots$ and $B(x_0, R) \subset \Omega$, where

$$R = \frac{2 + 5\vartheta - \sqrt{(2+5\vartheta)^2 - 48\vartheta}}{12L_0\beta}.$$

From (Q3), it follows that R is a real number.

Under the last three conditions, we can now guarantee the semilocal convergence of Newton's method in the following result [43].

Corollary 7.5 *Let* $F : \Omega \subseteq X \to Y$ *be a continuously Fréchet differentiable operator defined on a nonempty open convex domain* Ω *of a Banach space* X *with values in a Banach space* Y. *Suppose that the condition* (Q1)-(Q2)-(Q3) *are satisfied. Then, the Newton sequence* $\{x_n\}$ *starting at* x_0 *converges to a solution* x^* *of* $F(x) = 0$. *Moreover,* $x_n, x^* \in \overline{B(x_0, R)}$ *and* x^* *is unique in* $B(x_0, r) \cap \Omega$, *where*

$$r = \frac{22 - 5\vartheta + \sqrt{(2+5\vartheta)^2 - 48\vartheta}}{12L_0\beta}.$$

As we have mentioned above, taking into account that $L_0 \leq L$ and Corollary 7.5, we demonstrate the following simple example that condition $L_0\beta\eta \leq 0.168081\ldots < \frac{1}{2}$

holds, but not $L\beta\eta \leq \frac{1}{2}$ and, as a consequence, new starting points for Newton's method are found. For this, we study the accessibility of Newton's method using the experimental form given by the region of accessibility.

Example 7.6 Consider the simple example given by the complex equation $d(z) = z^{15} - 2 = 0$, and let us analyze the region of accessibility of solution $z^* = 1.0472\ldots$ when it is approximated by Newton's method [35].

Since an operator is center Lipschitz continuous in every point of a domain if it is Lipschitz continuous in all the domain, we have $L_0 \leq L$, where L is the Lipschitz constant and L_0 is the center Lipschitz constant in every point. Therefore, we can improve the Newton-Kantorovich Theorem 2.1 using Corollary 7.5, since we can choose a constant L_0 for each point of the domain considered by Theorem 2.1 of Newton-Kantorovich. For this, we take the ball $B\left(0, \frac{3}{2}\right)$ as the domain of the function $d(z)$. It is clear that $d'(z)$ is center Lipschitz in every point of $B\left(0, \frac{3}{2}\right)$ if $d'(z)$ is Lipschitz in $B\left(0, \frac{3}{2}\right)$. Next, we compare the regions of accessibility of $z^* = 1.0472\ldots$ when the Newton-Kantorovich Theorem 2.1 and Corollary 7.5 are used. To paint the regions of accessibility, we color all the points $z_0 \in B\left(0, \frac{3}{2}\right)$ that satisfy the condition $L\beta\eta \leq \frac{1}{2}$ with $L = 40870.1$ of the Newton-Kantorovich Theorem 2.1 (the yellow region in Fig. 7.2) and the condition $\vartheta =$

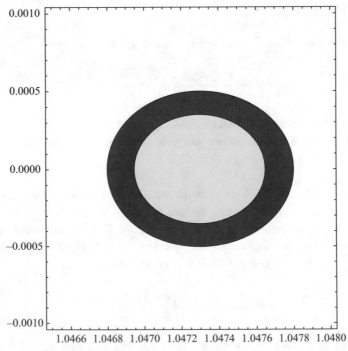

Fig. 7.2 Regions of accessibility of the root $x^* = 1.0472\ldots$

$L_0 \beta \eta \leq \frac{14-4\sqrt{6}}{25} = 0.168081\ldots$ with $L_0 = 15 \left(\sum_{i=0}^{13} \left(\frac{3}{2} \right)^i \|z_0\|^{13-i} \right)$ of Corollary 7.5 for each L_0 (blue and yellow regions in Fig. 7.2). Observe that Corollary 7.5 gives a better result, so that we can choose starting points for Newton's method for which the semilocal convergence of the method is guaranteed in the blue region.

In the next example, we go back to the nonlinear elliptic equation given in the introduction of the chapter to illustrate the preceding discussion.

Example 7.7 Consider the Eq. (7.7),

$$u_{xx} + u_{yy} = u^{\frac{3}{2}},$$

satisfied in the interior of the square $0 \leq x, y \leq 1$ in $\mathbb{R}^2$ and suppose $u(x, y) > 0$ is given and continuous on the boundary of the square and satisfies the boundary conditions in (7.14).

Applying the method of finite differences applied to the foregoing problem, as in the introduction of the chapter, yields a finite system of nonlinear equations. If we choose $N = M = 4$, then $\mathfrak{s} = \frac{1}{5}$, $\lambda = 1$ and the number of equations is $\mathfrak{m} = NM = 16$. For the problem (7.7)–(7.14), we obtain the system (7.6) with $q(\mathbf{x}) = \left(x_1^{\frac{3}{2}}, x_2^{\frac{3}{2}}, \ldots, x_{16}^{\frac{3}{2}} \right)^T$ and $\mathbf{v}$ as in (7.15).

Then, taking into account that $\mathbf{x}_0 = \mathbf{1} = (1, 1, \ldots, 1)^T \in \mathbb{R}^{16}$, we obtain

$$\|\mathbb{F}'(\mathbf{x}) - \mathbb{F}'(\mathbf{x}_0)\| \leq L_0 \|\mathbf{x} - \mathbf{x}_0\|$$

with $L_0 = \frac{3}{50}$. Moreover, $\beta = 1.5385\ldots$, $\eta = 0.8549\ldots$ and $\vartheta = L_0 \beta \eta = 0.0789\ldots \leq \frac{14-4\sqrt{6}}{25} = 0.168081\ldots$, so that the convergence of Newton's method to a solution $\mathbf{x}^*$ of the system $\mathbb{F}(\mathbf{x}) = 0$ with $\mathbb{F}$ given in (7.6) and $\Psi(u) = u^{\frac{3}{2}}$ is guaranteed by Corollary 7.5. Moreover, the domains of existence and uniqueness of the solution are respectively

$$\{\mathbf{v} \in \mathbb{R}^{16} : \|\mathbf{v} - \mathbf{1}\| \leq 0.9024\ldots\} \quad \text{and} \quad \{\mathbf{v} \in \mathbb{R}^{16} : \|\mathbf{v} - \mathbf{1}\| < 18.2446\ldots\}.$$

Table 7.3 Approximation of the solution $\mathbf{x}^*$ of $\mathbb{F}(\mathbf{x}) = 0$ with $\mathbb{F}$ given in (7.6) and $\Psi(u) = u^{\frac{3}{2}}$

i	x_i^*	i	x_i^*	i	x_i^*	i	x_i^*
1	0.98042539...	5	1.10026644...	9	1.29570719...	13	1.59164176...
2	1.10026644...	6	1.25109753...	10	1.42991628...	14	1.67118063...
3	1.29570719...	7	1.42991628...	11	1.57007502...	15	1.74958067...
4	1.59164176...	8	1.67118063...	12	1.74958067...	16	1.84963506...

Table 7.4 Absolute errors obtained by Newton's method and $\{\|\mathbb{F}(\mathbf{x}_n)\|\}$

n	$\|\mathbf{x}_n - \mathbf{x}^*\|$	$\|\mathbb{F}(\mathbf{x}_n)\|$
0	$8.4963\ldots \times 10^{-1}$	1.96
1	$6.5858\ldots \times 10^{-3}$	$9.7589\ldots \times 10^{-3}$
2	$5.0072\ldots \times 10^{-7}$	$5.1842\ldots \times 10^{-7}$
3	$2.5445\ldots \times 10^{-15}$	$3.0014\ldots \times 10^{-15}$

The solution $\mathbf{x}^*$ given in Table 7.3 is reached after four iterations of Newton's method. In Table 7.4 we show the errors $\|\mathbf{x}_n - \mathbf{x}^*\|$ and the sequence $\{\|\mathbb{F}(\mathbf{x}_n)\|\}$. Notice that the vector shown in Table 7.3 is a good approximation of a solution.

Remark 7.8 Finally, we note that another particular case of operators with center ω-Lipschitz first Fréchet derivative is when F' is center Hölder continuous, which is obtained when $\omega_0(z) = K_0 z^p$ and $h_0(t) = t^p$ with $p \in [0, 1]$ in (P2).

Using Center ω-Lipschitz Conditions for the First Derivative at Auxiliary Points

8

Recall that the domain of parameters associated with the Newton-Kantorovich Theorem 2.1 is

$$D = \left\{ (x, y) \in \mathbb{R}_+^2 : xy \leq \frac{1}{2} \right\},$$

so that the semilocal convergence of Newton's method is guaranteed for any starting point x_0 provided the values of x and y are such that $xy = L\beta\eta \leq \frac{1}{2}$, where the parameters L, β and η are fixed by the following conditions:

(A1)   There exists the operator $\Gamma_0 = [F'(x_0)]^{-1} \in \mathcal{L}(Y, X)$, for some $x_0 \in \Omega$, with $\|\Gamma_0\| \leq \beta$ and $\|\Gamma_0 F(x_0)\| \leq \eta$.

(A2)   There exists a constant $L \geq 0$ such that $\|F'(x) - F'(y)\| \leq L\|x - y\|$ for $x, y \in \Omega$.

(A3) $a_0 = L\beta\eta \leq \frac{1}{2}$ and $B(x_0, R) \subset \Omega$, where $R = \frac{2(1-a_0)}{2-3a_0}\eta$.

On the other hand, we can modify condition (A2) by centering:

(B2)   There exist $\widetilde{x} \in \Omega$ and $\widetilde{L} \in \mathbb{R}_+$ such that $\|F'(x) - F'(\widetilde{x})\| \leq \widetilde{L}\|x - \widetilde{x}\|$, $x \in \Omega$.

The modification (B2) of (A2) requires us to modify condition (A3). This modification is then more restrictive than condition (A3), since condition (B2) is milder than condition (A2). In the results known so far, in which condition (B2) is imposed to the operator F' instead of (A2), one usually takes $\widetilde{x} = x_0$, as we can see in Sect. 7.3, where the condition

(A3b) $\widetilde{L}\beta\eta \leq \frac{14-4\sqrt{6}}{25} = 0.1680816\ldots$

© The Editor(s) (if applicable) and The Author(s), under exclusive licence to Springer Nature Switzerland AG 2020

J. A. Ezquerro Fernández, M. Á. Hernández Verón, *Mild Differentiability Conditions for Newton's Method in Banach Spaces*, Frontiers in Mathematics, https://doi.org/10.1007/978-3-030-48702-7_8

is required, instead of (A3), to guarantee the semilocal convergence of Newton's method. So, the semilocal convergence of Newton's method is established in Corollary 7.5 under conditions (A1)-(B2)-(A3b).

Let us emphisize that Corollary 7.5 guarantees the semilocal convergence of Newton's method only when the point $\widetilde{x}$ where the operator F' is center Lipschitz is taken as the initial point x_0. Obviously, this choice is very restrictive. For example, if the point $\widetilde{x}$ is far from the solution x^* of $F(x) = 0$, condition (A3b) of the theorem is hardly verifiable.

In addition, given $\widetilde{x} = x_0 \in \Omega$, in order to guarantee the semilocal convergence of Newton's method based on Corollary 7.5, the parameters $\widetilde{L}$, β and η associated with x_0 must satisfies condition (A3b) and we cannot specify a domain of parameters, since we only can check if the starting point $\widetilde{x} = x_0$ verifies condition (A3b). In contrast to the Newton-Kantorovich Theorem 2.1, we cannot find other different starting points of $\widetilde{x} = x_0$ that guarantee the semilocal convergence of Newton's method, since the constant $\widetilde{L}$ is associated with a single point, $\widetilde{x}$, while the constant L in Theorem 2.1 is associated with any point of the domain Ω. In view of this, one of the aims of this chapter is to obtain semilocal convergence results under center conditions of type (B2), that have a domain of parameters associated along the same way as in Theorem 2.1, and unlike what happens with Corollary 7.5.

Notice that, at the beginning, condition (B2) has two advantages over condition (A2). First, if an operator satisfies (A2), it also satisfies (B2) and then $\widetilde{L} \leq L$, and, for certain operators, values of $\widetilde{L}$ such that the starting point $x_0 = \widetilde{x}$ satisfies (A3b), but not (A3), can be found. This increases the choices of good starting points for Newton's method. Second, there exist operators that satisfy (B2), but not (A2), as we have seen in Chap. 7 for $\widetilde{x} = x_0$.

In this chapter, we obtain new semilocal convergence results for Newton's method under center conditions on F' of type (B2) without requiring that the starting point x_0 coincides with the point $\widetilde{x}$. The semilocal convergence results known until now under center conditions of type (B2) have been proved under the assumption that $x_0 = \widetilde{x}$ is a good starting point for Newton's method [10, 11, 15, 43–46], but they do not allow us to locate other possible starting points. However, we see there that we can find a domain of starting points for Newton's method under condition (B2), since we show that the parameter $\widetilde{L}$ is associated with any point of the domain (as it happens in Theorem 2.1), so that we can find new starting points for Newton's method under a variant of condition (A3). We will even see that we can find starting points for Newton's method when $x_0 = \widetilde{x}$ is not a good starting point.

We emphasize that the technique used to prove the semilocal convergence of Newton's method is based on recurrence relations and using this technique we generalize all the results obtained previously by other authors [15, 43, 45].

To give sufficient generality to our study, we will consider all Lipschitz-continuity type conditions considered by other authors in connection with the semilocal convergence of

Newton's method. For instance, it is also common to require that F' is Hölder continuous in Ω instead of Lipschitz continuous; that is:

(A2b)   There exist two constants $K \geq 0$ and $p \in [0, 1]$ such that $\|F'(x) - F'(y)\| \leq K\|x - y\|^p$, $x, y \in \Omega$.

This condition is used in [48, 58, 72].

The case where F' is given by a combination of operators that are Lipschitz or Hölder continuous in Ω was also considered (see [31]); that is:

(A2c)   There exist $2m$ constants $K_1, K_2, \ldots, K_m \geq 0$ and $p_1, p_2, \ldots, p_m \in [0, 1]$ such that $\|F'(x) - F'(y)\| \leq \sum_{i=1}^{m} K_i \|x - y\|^{p_i}$, $x, y \in \Omega$.

To give sufficient generality to conditions (A2), (A2b) and (A2c) on the first Fréchet derivative F', we introduce the following condition:

(A2d)   There exist two continuous and nondecreasing functions $\omega : [0, +\infty) \to \mathbb{R}$ and $h : [0, 1] \to \mathbb{R}$ such that $\omega(0) \geq 0$, $\omega(tz) \leq h(t)\omega(z)$, with $t \in [0, 1]$, $z \in [0, \infty)$ and $\|F'(x) - F'(y)\| \leq \omega(\|x - y\|)$, $x, y \in \Omega$.

Obviously, we obtain (A2) if $\omega(z) = Lz$ and $h(t) = t$, (A2b) if $\omega(z) = Kz^p$ and $h(t) = t^p$, and (A2c) if $\omega(z) = \sum_{i=1}^{m} K_i z^{p_i}$ and $h(t) = t^p$, where $p = \min_{i=1,2,\ldots,m}\{p_i\}$ and $p_i \in [0, 1]$, for all $i = 1, 2, \ldots, m$. Moreover, note that condition $\omega(tz) \leq h(t)\omega(z)$, for $t \in [0, 1]$, is not restrictive, since h always exists, such that $h(t) = 1$, for $t \in [0, 1]$, because ω is a nondecreasing function.

Since our interest in this chapter is to widen the above conditions, we will assume the operator F' satisfies the following center condition:

(B2b)   There exist $\widetilde{x} \in \Omega$ and two continuous and nondecreasing functions $\widetilde{\omega} : [0, +\infty) \to \mathbb{R}$ and $\widetilde{h} : [0, 1] \to \mathbb{R}$ such that $\widetilde{\omega}(0) \geq 0$, $\widetilde{\omega}(tz) \leq \widetilde{h}(t)\widetilde{\omega}(z)$, with $t \in [0, 1]$, $z \in [0, \infty)$ and $\|F'(x) - F'(\widetilde{x})\| \leq \widetilde{\omega}(\|x - \widetilde{x}\|)$, $x \in \Omega$.

Notice that for suitable choices of the functions $\widetilde{\omega}$ and $\widetilde{h}$ condition (B2b) reduces to the cases where F' is center Lipschitz, F' is center Hölder or the corresponding center condition (A2c) in Ω is satisfied.

Next, in Sect. 8.1, we give the main semilocal convergence result for Newton's method under condition (B2b). In Sect. 8.2, we analyse three particular cases: F' is Lipschitz continuous in Ω, F' is Hölder continuous in Ω, and F' is a linear combination of the Lipschitz and the Hölder cases. An interesting analysis of the domain of parameters associated with the semilocal convergence result in the Lipschitz case is given, which clearly demonstrates the influence of center conditions on the improvement of the domain of starting points. In addition, we illustrate the Lipschitz case with two examples of systems of nonlinear equations that arise from the discretization of nonlinear elliptic equations.

8.1   Convergence Analysis

In this section, we establish a semilocal convergence result for Newton's method under a center condition of type (B2b). In particular, we assume that the following conditions [35] are satisfied:

(C1) There exists $\widetilde{x} \in \Omega$ such that $\|x_0 - \widetilde{x}\| = \alpha$, where $x_0 \in \Omega$.
(C2) There exists the operator $\Gamma_0 = [F'(x_0)]^{-1} \in \mathcal{L}(Y, X)$, for some $x_0 \in \Omega$, with $\|\Gamma_0\| \leq \beta$ and $\|\Gamma_0 F(x_0)\| \leq \eta$.
(C3) There exists a nondecreasing continuous function $\widetilde{\omega} : [0, +\infty) \to \mathbb{R}$ such that $\|F'(x) - F'(\widetilde{x})\| \leq \widetilde{\omega}(\|x - \widetilde{x}\|)$ for $x \in \Omega$ and $\omega(0) = 0$.

First of all, we give a technical lemma which is used later and whose proof, based on the algorithm of Newton's method and Taylor's series, is immediate.

Lemma 8.1 *Let $\{x_n\}$ be the sequence given by Newton's method. Then, for $n \in \mathbb{N}$,*

$$F(x_n) = \int_0^1 \left[\left(F'(x_{n-1} + \tau(x_n - x_{n-1})) - F'(\widetilde{x}) \right) + \left(F'(\widetilde{x}) - F'(x_{n-1}) \right) \right] d\tau \, (x_n - x_{n-1}).$$

Next, Lemma 8.1 and (C3) yield

$$\|F(x_n)\| \leq \int_0^1 [\widetilde{\omega}(\|x_{n-1} + \tau(x_n - x_{n-1}) - x_0\| + \|x_0 - \widetilde{x}\|)$$

$$+ \widetilde{\omega}(\|x_{n-1} - x_0\| + \|x_0 - \widetilde{x}\|)] \, d\tau \, \|x_n - x_{n-1}\|, \qquad (8.1)$$

for all $x_{n-1}, x_n, \widetilde{x} \in \Omega$.

Next, we establish the existence of the needed inverse operators.

Lemma 8.2 *Let $\{x_n\}$ be the sequence given by Newton's method. If $x_n \in B(x_0, \rho)$, for all $n \in \mathbb{N}$, and $\beta \left(\widetilde{\omega}(\alpha) + \widetilde{\omega}(\alpha + \rho) \right) < 1$, then*

(a) *The operator $\widetilde{\Gamma} = [F'(\widetilde{x})]^{-1}$ exists and $\|\widetilde{\Gamma}\| \leq \dfrac{\beta}{1 - \beta\widetilde{\omega}(\alpha)}$.*

(b) *The operator $\Gamma_n = [F'(x_n)]^{-1}$ exists and $\|\Gamma_n\| \leq \dfrac{\beta}{1 - \beta(\widetilde{\omega}(\alpha) + \widetilde{\omega}(\alpha + \rho))}$.*

Proof Item (a) follows from the estimates

$$\|I - \Gamma_0 F'(\widetilde{x})\| \leq \|\Gamma_0\| \|F'(x_0) - F'(\widetilde{x})\| \leq \beta\widetilde{\omega}(\|x_0 - \widetilde{x}\|) \leq \beta\widetilde{\omega}(\alpha) < 1$$

and the Banach lemma on invertible operators.

Similarly, item (b) follows from the estimates

$$\|I - \widetilde{\Gamma} F'(x_n)\| \leq \|\widetilde{\Gamma}\| \|F'(\widetilde{x}) - F'(x_n)\|$$

$$\leq \frac{\beta}{1 - \beta\widetilde{\omega}(\alpha)} \widetilde{\omega}(\|x_n - \widetilde{x}\|)$$

$$\leq \frac{\beta}{1 - \beta\widetilde{\omega}(\alpha)} \widetilde{\omega}(\|x_n - x_0\| + \|x_0 - \widetilde{x}\|)$$

$$\leq \frac{\beta\widetilde{\omega}(\alpha + \rho)}{1 - \beta\widetilde{\omega}(\alpha)}$$

$$< 1$$

and the Banach lemma on invertible operators. ∎

Now we present a new semilocal convergence result for Newton's method under conditions (C1)-(C2)-(C3).

Theorem 8.3 *Let $F : \Omega \subseteq X \to Y$ be a continuously Fréchet differentiable operator defined on a nonempty open convex domain Ω of a Banach space X with values in a Banach space Y. Suppose that the conditions (C1)-(C2)-(C3) are satisfied. If the equation*

$$t = \frac{1 - 2\beta\widetilde{\omega}(\alpha + t)}{1 - \beta(\widetilde{\omega}(\alpha) + 3\widetilde{\omega}(\alpha + t))} \eta \tag{8.2}$$

has at least one positive root and the smallest positive root, denoted by R, satisfies

$$\beta(\widetilde{\omega}(\alpha) + 3\widetilde{\omega}(\alpha + R)) < 1 \tag{8.3}$$

and $B(x_0, R) \subset \Omega$, then the Newton sequence $\{x_n\}$ starting at x_0 converges to a solution x^* of the equation $F(x) = 0$ and $x_n, x^* \in \overline{B(x_0, R)}$, for all $n = 0, 1, 2, \ldots$ Moreover, the solution x^* is unique in $B(x_0, r) \cap \Omega$, where r is the smallest positive root of the equation

$$\frac{\beta(\psi(r) - \psi(R))}{(1 - \beta\widetilde{\omega}(\alpha))(r - R)} = 1,$$

with $\psi(s) = \displaystyle\int_0^{\alpha+s} \widetilde{\omega}(t)\, dt$ and $s > 0$.

Proof We begin by proving that the sequence generated by Newton's method is well defined and $x_n \in B(x_0, R) \subset \Omega$ for all $n \in \mathbb{N}$.

From (C2), it follows that x_1 is well defined and $x_1 \in B(x_0, R)$, since $\eta < R$.

Next, (8.3) and item (b) of Lemma 8.2 ensures that the operator Γ_1 exists and

$$\|\Gamma_1\| \leq \frac{\beta}{1 - \beta(\widetilde{\omega}(\alpha) + \widetilde{\omega}(\alpha + R))}.$$

Further, inequality (8.1) with $n = 1$ shows that

$$\|F(x_1)\| \leq \int_0^1 (\widetilde{\omega}(\|x_0 - \widetilde{x}\| + \tau\|x_1 - x_0\|) + \widetilde{\omega}(\|x_0 - \widetilde{x}\|)) \, d\tau \|x_1 - x_0\|$$

$$\leq (\widetilde{\omega}(\alpha + R) + \widetilde{\omega}(\alpha)) \|x_1 - x_0\|.$$

Then

$$\|x_2 - x_1\| \leq \|\Gamma_1\| \|F(x_1)\| \leq P\|x_1 - x_0\|,$$

$$\|x_2 - x_0\| \leq \|x_2 - x_1\| + \|x_1 - x_0\| \leq (1 + P)\|x_1 - x_0\| \leq (1 + P)\eta < R,$$

where $P = \dfrac{\beta(\widetilde{\omega}(\alpha) + \widetilde{\omega}(\alpha + R))}{1 - \beta(\widetilde{\omega}(\alpha) + \widetilde{\omega}(\alpha + R))}$. Thus, $x_2 \in B(x_0, R)$.

Again, (8.3) and item (b) of Lemma 8.2 ensures that the operator Γ_2 exists and

$$\|\Gamma_2\| \leq \frac{\beta}{1 - \beta(\widetilde{\omega}(\alpha) + \widetilde{\omega}(\alpha + R))}.$$

Continuing, inequality (8.1) and the fact that $x_1 + \tau(x_2 - x_1) \in B(x_0, R)$ for $\tau \in [0, 1]$ yield

$$\|F(x_2)\| \leq 2\widetilde{\omega}(\alpha + R)\|x_2 - x_1\|.$$

Hence, by (8.3),

$$\|x_3 - x_2\| \leq Q\|x_2 - x_1\|,$$

$$\|x_3 - x_0\| \leq \|x_3 - x_2\| + \|x_2 - x_0\|$$

$$\leq Q\|x_2 - x_1\| + (1 + P)\|x_1 - x_0\|$$

$$\leq (1 + P + PQ)\|x_1 - x_0\|$$

$$< \left(1 + \frac{P}{1 - Q}\right)\eta$$

$$= R,$$

where $Q = \dfrac{2\beta(\widetilde{\omega}(\alpha + R))}{1 - \beta(\widetilde{\omega}(\alpha) + \widetilde{\omega}(\alpha + R))}$. Thus, $x_3 \in B(x_0, R)$.

Now we use induction. Namely, we assume that

- the operator Γ_i exists and $\|\Gamma_i\| \leq \dfrac{\beta}{1 - \beta(\widetilde{\omega}(\alpha) + \widetilde{\omega}(\alpha + R))}$;
- $\|F(x_i)\| \leq 2\widetilde{\omega}(\alpha + R)\|x_i - x_{i-1}\|$;
- $\|x_{i+1} - x_i\| \leq Q\|x_i - x_{i-1}\|$;
- $\|x_{i+1} - x_0\| \leq (1 + P + PQ + \cdots + PQ^{i-1})\eta < R$,

for all $i = 2, 3, \ldots, n$, and prove that these four inequalities hold for $i = n + 1$.

As above, (8.3) and item (b) of Lemma 8.2 ensures that Γ_{n+1} exists and

$$\|\Gamma_{n+1}\| \leq \frac{\beta}{1 - \beta(\widetilde{\omega}(\alpha) + \widetilde{\omega}(\alpha + R))}.$$

Then, from (8.1) and the fact that $x_n + \tau(x_{n+1} - x_n) \in B(x_0, R)$ for $\tau \in [0, 1]$, we obtain

$$\|F(x_{n+1})\| \leq 2\widetilde{\omega}(\alpha + R)\|x_{n+1} - x_n\|.$$

Therefore

$$\|x_{n+2} - x_{n+1}\| \leq Q\|x_{n+1} - x_n\|,$$

$$\|x_{n+2} - x_0\| \leq \|x_{n+2} - x_{n+1}\| + \|x_{n+1} - x_0\|$$

$$\leq PQ^n\eta + \left(1 + P + PQ + \cdots + PQ^{n-1}\right)\eta$$

$$< R.$$

Thus, $x_{n+2} \in B(x_0, R)$. We have shown that $x_n \in B(x_0, R)$, for all $n \in \mathbb{N}$, so the sequence generated by Newton's method $\{x_n\}$ is well defined. Let us verify that $\{x_n\}$ is a Cauchy sequence. Indeed,

$$\|x_{n+j} - x_n\| \leq \sum_{i=1}^{j} \|x_{n+i} - x_{n+i-1}\| \leq \sum_{i=1}^{j} PQ^{n+i-2}\|x_1 - x_0\| < \frac{PQ^{n-1}}{1 - Q}\|x_1 - x_0\|,$$

for $j \geq 1$, and since $Q < 1$, it is clear that $\{x_n\}$ is a Cauchy sequence. Consequently, $\{x_n\}$ converges. Now, if $\lim\limits_{n \to \infty} x_n = x^*$, it follows that $F(x^*) = 0$, by the continuity of the operator F; indeed

$$\|F(x_n)\| \leq 2\widetilde{\omega}(\alpha + R)\|x_n - x_{n-1}\|$$

and $\|x_n - x_{n-1}\| \to 0$ as $n \to \infty$.

Finally, to prove the uniqueness of the solution x^*, suppose that y^* is another solution of $F(x) = 0$ in $B(x_0, r) \cap \Omega$. Then, from the approximation

$$0 = F(y^*) - F(x^*) = \int_{x^*}^{y^*} F'(x)dx = \int_0^1 F'(x^* + \tau(y^* - x^*)) \, d\tau (y^* - x^*),$$

it follows that $x^* = y^*$ if we can show that the operator $J = \int_0^1 F'(x^* + \tau(y^* - x^*)) \, d\tau$ is invertible. So, since

$$\|I - \widetilde{\Gamma}\| \leq \|\widetilde{\Gamma}\| \int_0^1 \|F'(\widetilde{x}) - F'(x^* + \tau(y^* - x^*))\| \, d\tau$$

$$\leq \|\widetilde{\Gamma}\| \int_0^1 \widetilde{\omega}\left((1 - \tau)\|x^* - \widetilde{x}\| + \tau\|y^* - \widetilde{x}\|\right) d\tau$$

$$< \|\widetilde{\Gamma}\| \int_0^1 \widetilde{\omega}\left(\alpha + R + \tau(r - R)\right) d\tau$$

$$\leq \frac{\beta}{(1 - \beta\widetilde{\omega}(\alpha))(r - R)} \int_{\alpha+R}^{\alpha+r} \widetilde{\omega}(t) \, dt$$

$$= 1,$$

the operator J^{-1} exists by the Banach lemma on invertible operators. The proof is complete. ∎

It is easy to see that the uniqueness of the solution is guaranteed in $B(x_0, R)$ if $\widetilde{\omega}(\alpha + R) + \widetilde{\omega}(\alpha) = \frac{1}{\beta}$.

Remark 8.4 As a first generalization of the semilocal convergence result obtained above, we notice that Theorem 8.3 reduces to Theorem 7.3 if $\widetilde{x} = x_0$.

8.2   Particular Cases and Applications to Elliptic Problems

Let us examine three different situations that one can encounter depending on the nature of the function $\widetilde{\omega}$ is. In particular, we will analyze the cases in which F' is Lipschitz continuous in Ω, F' is Hölder continuous in Ω, and F' is a linear combination of the Lipschitz and Hölder cases (which is called general case).

8.2.1 The Lipschitz Case

We begin by studying the simpler case [36]: F' is center Lipschitz at $\tilde{x}$. In this case, from conditions (A2b) and (B2b), we can consider the particular case of (B2b) given by $\tilde{\omega}(z) = \tilde{L}z$, where $\tilde{L}$ is constant; then F' is center Lipschitz continuous in Ω. In this particular situation we know from the results obtained above that to guarantee the convergence of Newton's method from Theorem 8.3, the following condition must be satisfied:

$$\tilde{L}\beta(4\alpha + 3R) < 1, \tag{8.4}$$

where R is the smallest positive real root of the equation

$$t = \frac{1 - \tilde{L}\beta\left(2\alpha + \frac{5}{2}t\right)}{1 - \tilde{L}\beta(4\alpha + 3t)}\,\eta. \tag{8.5}$$

In addition, the following result can be established, which reduces to Corollary 7.5 if $\alpha = 0$.

Corollary 8.5 *Let $F : \Omega \subseteq X \rightarrow Y$ be a continuously Fréchet differentiable operator defined on a nonempty open convex domain Ω of a Banach space X with values in a Banach space Y. Suppose that the conditions (C1)-(C2)-(C3) with $\tilde{\omega}(z) = \tilde{L}z$, where $\tilde{L}$ is constant, are satisfied. If Eq. (8.5) has at least one positive root and its smallest positive root, denoted by R, satisfies condition (8.4) and $B(x_0, R) \subset \Omega$, then the Newton sequence $\{x_n\}$ starting at x_0 converges to a solution x^* of the equation $F(x) = 0$ and $x_n, x^* \in \overline{B(x_0, R)}$, for all $n = 0, 1, 2, \dots$ Moreover, the solution x^* is unique in $B(x_0, r) \cap \Omega$, where $r = \frac{2}{\tilde{L}\beta} - (4\alpha + R)$, provided that $R < \frac{2(1 - 2\tilde{L}\alpha\beta)}{\tilde{L}\beta}$.*

Remark 8.6 Note that Corollary 8.5 reduces to Corollary 7.5 if $\tilde{x} = x_0$.

8.2.2 Domain of Parameters

Now that we have established the semilocal convergence of Newton's method, we turn to the domain of parameters associated with Corollary 8.5.

To this end, we transform the Eq. (8.5) into the equivalent quadratic equation

$$6\tilde{L}\beta t^2 + \left(\tilde{L}\beta(8\alpha - 5\eta) - 2\right)t + 2(1 - 2\tilde{L}\alpha\beta)\eta = 0, \tag{8.6}$$

and determine when it has positive roots in the following lemma.

Lemma 8.7 *If*

$$(1 - 2\widetilde{L}\alpha\beta)\eta > 0 \tag{8.7}$$

and

$$\widetilde{L}\beta(8\alpha - 5\eta) - 2 + \sqrt{48\widetilde{L}\beta\eta(1 - 2\widetilde{L}\alpha\beta)} > 0, \tag{8.8}$$

then Eq. (8.6) has two positive roots. Moreover, if we denote the smallest root by R, i.e.,

$$R = \frac{1}{12\widetilde{L}\beta}\left(2 - \widetilde{L}\beta(8\alpha - 5\eta) - \sqrt{\Delta}\right), \tag{8.9}$$

where $\Delta = \left(\widetilde{L}\beta(8\alpha - 5\eta) - 2\right)^2 - 48\widetilde{L}\beta\eta\left(1 - 2\widetilde{L}\alpha\beta\right)$, then condition (8.4) of Corollary 8.5 is satisfied if

$$8\widetilde{L}\alpha\beta + 5\widetilde{L}\beta\eta - 2 - \sqrt{(\widetilde{L}\beta(8\alpha - 5\eta) - 2)^2 + 48\widetilde{L}\beta\eta(2\widetilde{L}\alpha\beta - 1)} > 0. \tag{8.10}$$

Note that we can also consider the existence of a double root in the previous lemma, by allowing non-strict inequalities.

We emphasize that we do not treat the case where the Eq. (8.6) has one positive root and one negative root because in this case the positive root does not satisfy condition (8.10).

In addition, we can state the following result, whose proof follows easily from Corollary 8.5.

Corollary 8.8 *Let $F : \Omega \subseteq X \rightarrow Y$ be a continuously Fréchet differentiable operator defined on a nonempty open convex domain Ω of a Banach space X with values in a Banach space Y. Suppose that the conditions (C1)-(C2)-(C3) with $\widetilde{\omega}(z) = \widetilde{L}z$, where $\widetilde{L}$ is constant, are satisfied. Suppose also that (8.7), (8.8) and (8.10) hold and $B(x_0, R) \subset D$, where R is defined in (8.9). Then the Newton sequence $\{x_n\}$ starting at x_0 converges to a solution x^* of the equation $F(x) = 0$ and $x_n, x^* \in \overline{B(x_0, R)}$, for all $n = 0, 1, 2, \dots$ Moreover, the solution x^* is unique in $B(x_0, r) \cap \Omega$, where $r = \frac{2}{\widetilde{L}\beta} - (4\alpha + R)$, provided that $R < \frac{2(1 - 2\widetilde{L}\alpha\beta)}{\widetilde{L}\beta}$.*

Now, let $x = \widetilde{L}\beta$ and $y = \eta$. Then define the domain of parameters associated with Corollary 8.8 for Newton's method as the region of the xy-plane whose points satisfy conditions (8.7), (8.8) and (8.10), so that the convergence of Newton's method is guaranteed by the hypotheses in Corollary 8.8. For this, we colour the values of the parameters that satisfy conditions (8.7), (8.8) and (8.10) in the xy-plane, see Fig. 8.1. Note that the initial conditions (C1)-(C2), imposed on the initial approximation x_0, define the parameters α, β and η, and the condition (C3) with $\widetilde{\omega}(z) = \widetilde{L}z$, imposed on the operator

Fig. 8.1 Domains of parameters of Newton's method associated with Corollary 8.8 when $\alpha = \frac{1}{2}, \frac{1}{3}, \frac{1}{5}, \frac{1}{10}, 0$ (green, red, yellow, blue and orange, respectively)

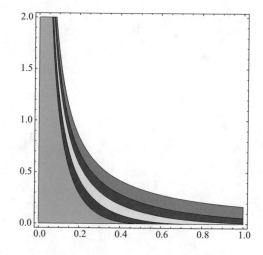

F, defines the fixed parameter $\widetilde{L}$. Figure 8.1 shows that, once $\|x_0 - \widetilde{x}\| = \alpha$ is fixed, we obtain a region of the xy-plane which contains the pair $(\widetilde{L}\beta, \eta)$, associated with $x_0 \in \Omega$, belongs, is obtained, so that Newton's method is convergent from the starting point x_0. Notice that this region of the xy-plane increases as the value of α decreases.

8.2.3 Region of Accessibility

As already mentioned, the region of accessibility represents, experimentally, the domain of starting points that satisfy the convergence conditions required by the iterative method that we want to apply for solving a particular equation.

To clarify the advantage of Corollary 8.5 compared to Corollary 7.5, we consider a simple academic example: the complex equation $d(z) = z^{3/2} - \frac{5}{4} = 0$, and analyze the region of accessibility of the root $z^* = 1.1604\ldots$ of $d(z) = 0$ when it is approximated by Newton's method. For this, we take the ball $B(0, 2)$ as the domain of the function $d(z)$ and analyse two situations.

In the first one, we consider $\widetilde{z} = 1.1$, $z_0 \in B(0, 2)$ and, to paint the region of accessibility, colour gray all the points z_0 that satisfy condition (8.4) of Corollary 8.5. So, $\widetilde{L} = 1.4301\ldots$ and, once $\alpha = |z_0 - \widetilde{z}| = |x_0 - 1.1|$ is fixed, we obtain the region of accessibility of $z^* = 1.1604\ldots$ shown in Fig. 8.2, where the white point is z^* and the red point is $\widetilde{z}$. Observe that we can choose $\widetilde{z} = 1.1$, the point where $d'(z)$ is center Lipschitz, as starting point for Newton's method.

In the second situation, we take $\widetilde{z} = 1$ and $z_0 \in B(0, 2)$, do the same as before to paint the region of accessibility, and obtain the region shown in Fig. 8.3, where the red point is now $\widetilde{z} = 1$ and $\widetilde{L} = \frac{3}{2}$. Observe that, in this case, we cannot choose $\widetilde{z} = 1$ as starting point for Newton's method.

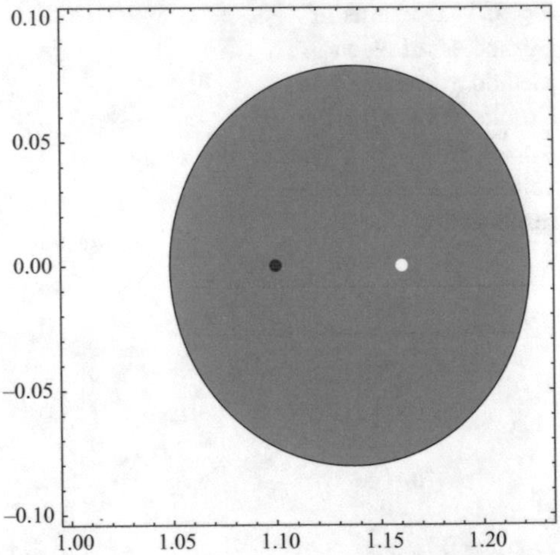

Fig. 8.2 Region of
accessibility of $z^* = 1.1604\ldots$
when $\widetilde{z} = 1.1$

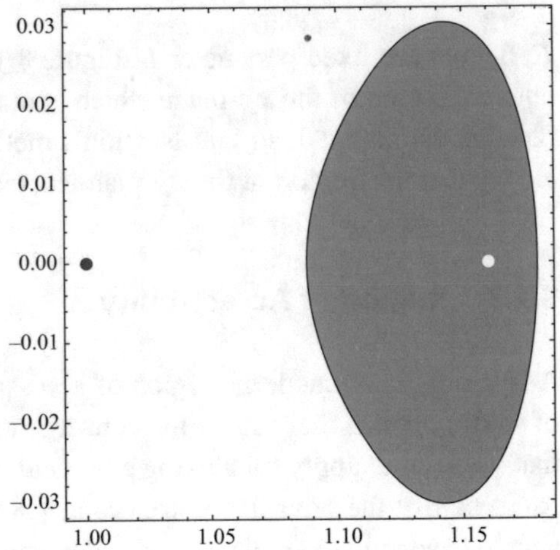

Fig. 8.3 Region of
accessibility of $z^* = 1.1604\ldots$
when $\widetilde{z} = 1$

Note that, in both situations, we can find starting points (points of gray regions) for Newton's method different from those where $d'(z)$ is center Lipschitz. As a consequence, the semilocal convergence of Newton's method is guaranteed for starting points different from $\widetilde{z} = 1$ (gray regions), since the points of the gray regions can be also chosen as starting points for which convergence is guaranteed by Corollary 8.5, while Corollary 7.5 can only guarantee convergence for the starting point $\widetilde{z}$.

Finally, we conjecture that as the point $\widetilde{z}$ where $d'(z)$ is center Lipschitz, moves far then from the solution $z^* = 1.1604\ldots$, the region of accessibility is smaller.

8.2.4 Applications

Let us consider two nonlinear elliptic problems of form (7.4)–(7.5), where the advantages revealed by the previous analysis are demonstrated.

Example 8.9 Consider again the nonlinear elliptic problem (7.7)–(7.14), i.e.,

$$u_{xx} + u_{yy} = u^{\frac{3}{2}}$$

with the boundary conditions

$$u(x, 0) = 2x^2 - x + 1, \ u(x, 1) = 2, \ 0 \leq x \leq 1,$$

$$u(0, y) = 2y^2 - y + 1, \ u(1, y) = 2, \ 0 \leq y \leq 1.$$

We have seen in Example 7.7 that

$$\|\mathbb{F}'(\mathbf{x}) - \mathbb{F}'(\mathbf{x}_0)\| \leq L_0 \|\mathbf{x} - \mathbf{x}_0\|$$

with $L_0 = \frac{3}{50}$ if $\mathbf{x}_0 = \mathbf{1} = (1, 1, \ldots, 1)^T \in \mathbb{R}^{16}$ and that the convergence of Newton's method to the solution $\mathbf{x}^*$ given in Table 7.3 is then guaranteed by Corollary 7.5, which also provides the domains of existence and uniqueness of the solution,

$$\{\mathbf{v} \in \mathbb{R}^{16} : \|\mathbf{v} - \mathbf{1}\| \leq 0.9024\ldots\} \quad \text{and} \quad \{\mathbf{v} \in \mathbb{R}^{16} : \|\mathbf{v} - \mathbf{1}\| < 18.2446\ldots\},$$

respectively.

The preceding analysis also shows that we can guarantee the semilocal convergence of Newton's method by Corollary 8.5, by choosing $\widetilde{\mathbf{x}} = \mathbf{1}$ as the point where the Lipschitz condition for $\mathbb{F}'$ is centered, and the starting point $\mathbf{x}_0 = \left(\frac{3}{2}, \frac{3}{2}, \ldots, \frac{3}{2} \right)^T$, which is different from $\widetilde{\mathbf{x}}$. In this case, $\alpha = \frac{1}{2}, \beta = 1.5123\ldots, \eta = 0.5180\ldots$ and the Eq. (8.5) reduces to the equation $(0.2722\ldots)t^2 - (0.9360\ldots)t + (0.4710\ldots) = 0$, which has two positive roots and the smallest one of which, $R = 0.6122\ldots$, satisfies condition (8.4) of Corollary 8.5, since $\widetilde{L}\beta(4\alpha + 3R) = 0.3481\ldots < 1$ with $\widetilde{L} = \frac{3}{50}$. Moreover, the domains of existence and uniqueness of the solution are respectively

$$\{\mathbf{v} \in \mathbb{R}^{16} : \|\mathbf{v} - \mathbf{1}\| \leq 0.6122\ldots\} \quad \text{and} \quad \{\mathbf{v} \in \mathbb{R}^{16} : \|\mathbf{v} - \mathbf{1}\| < 19.4291\ldots\}.$$

In this example we can choose as starting points for Newton's method points different from the one where the center Lipschitz condition for the first Fréchet derivative of the operator involved holds.

Example 8.10 Next, consider the nonlinear elliptic equation

$$u_{xx} + u_{yy} = u^3$$

in the interior of the square $0 \leq x, y \leq 1$ in $\mathbb{R}^2$, and that $u(x, y) > 0$ is given and continuous on the boundary of the square given and satisfies the boundary conditions in (7.14).

Next, we use the method of finite differences applied to problem (7.4)–(7.5) in the introduction of Chap. 7. If we choose $N = M = 4$, then $\mathfrak{s} = \frac{1}{5}$, $\lambda = 1$ and we obtain the system $\mathbb{F}(\mathbf{x}) = 0$, where $\mathbb{F}$ is given by (7.6),

$$\mathbb{F}(\mathbf{x}) = A\mathbf{x} + \mathfrak{s}^2 q(\mathbf{x}) - \mathbf{v}, \qquad \mathbb{F} : \mathbb{R}^m \longrightarrow \mathbb{R}^m,$$

with $m = NM = 16$, $q(\mathbf{x}) = \left(x_1^3, x_2^3, \ldots, x_{16}^3\right)^T$ and $\mathbf{v}$ as in (7.15).

Next, we observe that a solution $\mathbf{x}^*$ of the system $\mathbb{F}(\mathbf{x}) = 0$ always satisfies

$$\|\mathbf{x}^*\| \leq \|A^{-1}\| \left(\|\mathbf{v}\| + \mathfrak{s}^2 \|q(\mathbf{x})\| \right),$$

since

$$\|\mathbf{x}^*\| - \frac{5}{3} \left(4 + \frac{1}{25} \|\mathbf{x}^*\|^3 \right) \leq 0,$$

where $\|A^{-1}\| = \frac{5}{3}$ and $\|\mathbf{v}\| = 4$. Then we consider the map

$$\mathbb{F} : \Lambda \subset \mathbb{R}^{16} \longrightarrow \mathbb{R}^{16} \qquad \text{with} \qquad \Lambda = \left\{ \mathbf{x} \in \mathbb{R}^{16} : \|\mathbf{x}\| < \frac{1}{6} \right\}.$$

One verifies that

$$\mathbb{F}'(\mathbf{x}) = A + 3\mathfrak{s}^2 \operatorname{diag} \left\{ x_1^2, x_2^2, \ldots, x_{16}^2 \right\},$$

$$\mathbb{F}'(\mathbf{x}) - \mathbb{F}'(\mathbf{y}) = 3\mathfrak{s}^2 \operatorname{diag} \left\{ x_1^2 - y_1^2, x_2^2 - y_2^2, \ldots, x_{16}^2 - y_{16}^2 \right\},$$

$$\|\mathbb{F}'(\mathbf{x}) - \mathbb{F}'(\mathbf{y})\| \leq 3\mathfrak{s}^2 \left(\left(\frac{1}{6}\right) + \|\mathbf{y}\| \right) \|\mathbf{x} - \mathbf{y}\|,$$

where $\mathbf{y} = (y_1, y_2, \ldots, y_8)^T$.

Then, we study two cases. In the first one, we choose $\mathbf{x}_0 = \mathbf{1} = (1, 1, \ldots, 1)^T$ and see that we can guarantee the semilocal convergence of Newton's method by Corollary 7.5, since $L_0 = 3\mathfrak{s}^2 \left(\frac{1}{6} + \|\mathbf{x}_0\| \right) = 0.14$, $\beta = 1.4280\ldots$, $\eta = 0.8242\ldots$ and $\vartheta = L_0 \beta \eta = $

$0.1647\ldots \leq \frac{14-4\sqrt{6}}{25} = 0.168081\ldots$ In this case the domains of existence and uniqueness of solution are respectively

$$\{v \in \mathbb{R}^{16} : \|v - 1\| \leq 1.0709\ldots\} \quad \text{and} \quad \{v \in \mathbb{R}^{16} : \|v - 1\| < 8.7204\ldots\}.$$

On the other hand, we can also guarantee the semilocal convergence of Newton's method by Corollary 8.5 by choosing $\widetilde{x} = 1$ as the point where the Lipschitz condition is centered for $\mathbb{F}'$, and the starting point $x_0 = \left(\frac{7}{5}, \frac{7}{5}, \ldots, \frac{7}{5}\right)^T$, which is different from $\widetilde{x}$. In this case, $\alpha = \frac{2}{5}$, $\beta = 1.2537\ldots$, $\eta = 0.4209\ldots$ and the Eq. (8.5) reduces to $(0.5265\ldots)t^2 - (0.9038\ldots)t + (0.3618\ldots) = 0$, which has two positive roots and the smallest positive root of which, $R = 0.6358\ldots$, satisfies condition (8.4) of Corollary 8.5, since $\widetilde{L}\beta(4\alpha + 3R) = 0.4561\ldots < 1$, where $\widetilde{L} = 3\kappa^2\left(\frac{1}{6} + \|\widetilde{x}\|\right) = 0.14$. In this case the domains of existence and uniqueness of solution are respectively

$$\{v \in \mathbb{R}^{16} : \|v - x_0\| \leq 0.6358\ldots\} \quad \text{and} \quad \{v \in \mathbb{R}^{16} : \|v - x_0\| < 9.1589\ldots\}.$$

In the second case, we see that we can even go beyond. If we choose the starting point $x_0 = \left(\frac{3}{5}, \frac{3}{5}, \ldots, \frac{3}{5}\right)^T$, we cannot guarantee the semilocal convergence of Newton's method by Corollary 7.5, since $L_0 = 0.092$, $\beta = 1.5724\ldots$, $\eta = 1.2753\ldots$ and $\vartheta = L_0\beta\eta = 0.1845\ldots > \frac{14-4\sqrt{6}}{25} = 0.168081\ldots$

However, we can guarantee the semilocal convergence of Newton's method by Corollary 8.5 by choosing $\widetilde{x} = \left(\frac{3}{5}, \frac{3}{5}, \ldots, \frac{3}{5}\right)^T$ as the point where the Lipschitz condition is centered for $\mathbb{F}'$, and taking the starting point $x_0 = (1.3, 1.3, \ldots, 1.3)^T$, which is different from $\widetilde{x}$. Then, $\widetilde{L} = 0.092$, $\alpha = 0.7$, $\beta = 1.2984\ldots$, $\eta = 0.4951\ldots$ and Eq. (8.5) reduces to $(0.3583\ldots)t^2 - (0.8133\ldots)t + (0.4123\ldots) = 0$, which has two positive roots, the smallest one of which, $R = 0.7643\ldots$, satisfies condition (8.4) of Corollary 8.5, since $\widetilde{L}\beta(4\alpha + 3R) = 0.6084\ldots < 1$. In addition, the domains of existence and uniqueness of a solution are respectively

$$\{v \in \mathbb{R}^{16} : \|v - x_0\| \leq 0.7643\ldots\} \quad \text{and} \quad \{v \in \mathbb{R}^{16} : \|v - x_0\| < 13.1778\ldots\}.$$

In Fig. 8.4, we can see graphically that $x_0 = (1.3, 1.3, \ldots, 1.3)^T$ is a good starting point for Newton's method, since its corresponding pair $(\widetilde{L}\beta, \eta) = (0.1194\ldots, 0.4951\ldots)$, the black point, lies in the domain of parameters associated with Corollary 8.5 (pink region) when $\alpha = 0.7$. Note that, in this case, we cannot choose $\left(\frac{3}{5}, \frac{3}{5}, \ldots, \frac{3}{5}\right)^T$ as the starting point x_0 of a convergent Newton sequence, since the conditions of Corollary 7.5 are not satisfied.

After that, we use Newton's method starting at $x_0 = (1.3, 1.3, \ldots, 1.3)^T$ to approximate a solution of the system $\mathbb{F}(x) = 0$ with $\mathbb{F}$ given in (7.6) and $\Psi(u) = u^3$. The

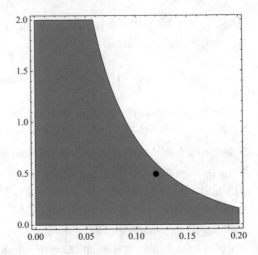

Fig. 8.4 Domain of parameters (pink region) associated with Corollary 8.5 when $\alpha = 0.7$ and Newton's method is applied to solve $\mathbb{F}(\mathbf{x}) = 0$ with $\mathbb{F}$ given in (7.6) and $\Psi(u) = u^3$

Table 8.1 Approximation of the solution $\mathbf{x}^*$ of $\mathbb{F}(\mathbf{x}) = 0$ with $\mathbb{F}$ given in (7.6) and $\Psi(u) = u^3$

i	x_i^*	i	x_i^*	i	x_i^*	i	x_i^*
1	$0.967514\ldots$	5	$1.073142\ldots$	9	$1.255308\ldots$	13	$1.547504\ldots$
2	$1.073142\ldots$	6	$1.199182\ldots$	10	$1.359712\ldots$	14	$1.602945\ldots$
3	$1.255308\ldots$	7	$1.359712\ldots$	11	$1.481965\ldots$	15	$1.669313\ldots$
4	$1.547504\ldots$	8	$1.602945\ldots$	12	$1.669313\ldots$	16	$1.778410\ldots$

Table 8.2 Absolute errors obtained by Newton's method and $\{\|\mathbb{F}(\mathbf{x}_n)\|\}$

n	$\|\mathbf{x}^* - \mathbf{x}_n\|$	$\|\mathbb{F}(\mathbf{x}_n)\|$
0	$4.7841\ldots \times 10^{-1}$	1.31212
1	$1.6727\ldots \times 10^{-2}$	$4.3100\ldots \times 10^{-2}$
2	$2.9183\ldots \times 10^{-5}$	$5.9902\ldots \times 10^{-5}$
3	$1.1857\ldots \times 10^{-10}$	$1.5253\ldots \times 10^{-10}$

approximation $\mathbf{x}^*$ given in Table 8.1 is reached after four iterations. In Table 8.2 we show the errors $\|\mathbf{x}^* - \mathbf{x}_n\|$ and the sequence $\{\|\mathbb{F}(\mathbf{x}_n)\|\}$. Notice that the vector in Table 8.1 is a good approximation of a solution of the nonlinear system under study.

As the last two examples demonstrate compared with Chap. 7, we can enlarge the domain of starting points for Newton's method by using center Lipschitz conditions for the first Fréchet derivative of the operator involved at auxiliary points.

8.2.5 The Hölder Case

Under the conditions (A2b) and (B2b), we can consider the particular case of (B2b) given by $\widetilde{\omega}(z) = \widetilde{K} z^p$ with $\widetilde{K}$ a constant and $p \in (0, 1]$. Then F' is center Hölder continuous in

Ω. In this particular situation, to guarantee the convergence of Newton's method by means of Theorem 8.3, the following condition must be satisfied:

$$\widetilde{K}\beta(4\alpha^p + 3R^p) < 1,$$

where R is the smallest positive root of the equation

$$t = \frac{1 - \widetilde{K}\beta\left(2\alpha^p + \dfrac{2+3p}{1+p}t^p\right)}{1 - \widetilde{K}\beta(4\alpha^p + 3t^p)}\eta.$$

In addition, we can state the following result.

Corollary 8.11 *Let $F : \Omega \subseteq X \to Y$ be a continuously Fréchet differentiable operator defined on a nonempty open convex domain Ω of a Banach space X with values in a Banach space Y. Suppose that the conditions (C1)-(C2)-(C3) with $\widetilde{\omega}(z) = \widetilde{K}z^p$, where $\widetilde{K}$ is constant and $p \in (0, 1]$, are satisfied. Suppose also that the Eq. (8.5) has at least one positive root and the smallest positive root, denoted by R, satisfies the condition (8.4) and $B(x_0, R) \subset \Omega$. Then the Newton sequence $\{x_n\}$ starting at x_0 converges to a solution x^* of the equation $F(x) = 0$ and $x_n, x^* \in \overline{B(x_0, R)}$, for all $n = 0, 1, 2, \ldots$ Moreover, the solution x^* is unique in $B(x_0, r) \cap \Omega$, where r is the smallest positive root of the equation $\frac{\widetilde{K}\beta((\alpha+r)^{1+p} - (\alpha+R)^{1+p})}{(1+p)(r-R)(1-\widetilde{K}\beta\alpha^p)} = 1$.*

8.2.6   General Case

Under the conditions (A2c) and (B2b), we can consider the particular case of (B2b) given by $\widetilde{\omega}(z) = \sum_{i=1}^{m} K_i^0 z^{p_i}$, where $p_i \in [0, 1]$, for all $i = 1, 2, \ldots, m$. Let us show that the following two properties hold:

(P1) $\widetilde{\omega}(u + v) = \sum_{i=1}^{m} K_i^0(u + v)^{p_i} \leq \sum_{i=1}^{m} K_i^0\left(u^{p_i} + v^{p_i}\right) = \widetilde{\omega}(u) + \widetilde{\omega}(v).$

(P2) $\widetilde{\omega}(\lambda u) = \sum_{i=1}^{m} K_i^0(\lambda u)^{p_i} \leq \lambda^p \sum_{i=1}^{m} K_i^0 u^{p_i} = \lambda^p \widetilde{\omega}(u), \lambda \in [0, 1].$

For property (P1), take $u > v$, so that $x = \frac{v}{u} \in (0, 1)$. Consider the auxiliary function $\varphi(x) = 1 + x^{p_i} - (1 + x)^{p_i}$; then $\varphi(0) = \varphi(1) = 0$ and $\varphi''(x) = p_i(p_i - 1)\left(x^{p_i - 2} - (1 + x)^{p_i - 2}\right) < 0$. Therefore,

$$(u + v)^{p_i} = u^{p_i}\left(1 + \left(\frac{v}{u}\right)\right)^{p_i} \leq u^{p_i}\left(1 + \left(\frac{v}{u}\right)^{p_i}\right) = u^{p_i} + v^{p_i}.$$

Property (P2) follows by simply observing that $\lambda^{p_i} \leq \lambda^p$ with $p = \min_{i=1,2,\ldots,m}\{p_i\}$, since $p_i \ln \lambda \leq p \ln \lambda$, for all $p_i \in [0, 1]$.

In view of the properties (P1)-(P2), Theorem 8.3 is then reduced to the following result.

Theorem 8.12 *Let $F : \Omega \subseteq X \to Y$ be a continuously Fréchet differentiable operator defined on a nonempty open convex domain Ω of a Banach space X with values in a Banach space Y. Suppose that the conditions (C1)-(C2)-(C3) and the properties (P1)-(P2) are satisfied. Suppose also that the equation*

$$t = \frac{1 - \beta \left(2\,\widetilde{\omega}(\alpha) + \left(3 - \frac{1}{1+p}\right)\widetilde{\omega}(t)\right)}{1 - \beta\,(4\,\widetilde{\omega}(\alpha) + 3\,\widetilde{\omega}(t))}\,\eta \tag{8.11}$$

has at least one positive root and its smallest positive root, denoted by R, satisfies

$$\beta\,(4\,\widetilde{\omega}(\alpha) + 3\,\widetilde{\omega}(R)) < 1 \tag{8.12}$$

and $B(x_0, R) \subset \Omega$. Then the Newton sequence $\{x_n\}$ starting at x_0 converges to a solution x^* of the equation $F(x) = 0$ and $x_n, x^* \in \overline{B(x_0, R)}$, for all $n = 0, 1, 2, \ldots$ Moreover, the solution x^* is unique in $B(x_0, r) \cap \Omega$, where r is the smallest positive root of the equation

$$\frac{\beta(\psi(r) - \psi(R))}{(1 - \beta\widetilde{\omega}(\alpha))(r - R)} = 1,$$

where $\psi(r) = \displaystyle\int_0^{\alpha+r} \widetilde{\omega}(t)\,dt$.

Proof From (C2) and (8.11), it is clear that $\|x_1 - x_0\| < R$, so that $x_1 \in B(x_0, R)$.
Next, (8.12), item (b) of Lemma 8.2 and properties (P1)–(P2) imply that

$$\|F(x_1)\| \le \int_0^1 (\widetilde{\omega}(\|x_0 - \widetilde{x}\|) + \widetilde{\omega}(\tau\,\|x_1 - x_0\|) + \widetilde{\omega}(\|x_0 - \widetilde{x}\|))\,d\tau\,\|x_1 - x_0\|$$

$$\le \left(2\,\widetilde{\omega}(\alpha) + \int_0^1 \tau^p\,\widetilde{\omega}(R)\,d\tau\right)\|x_1 - x_0\|$$

$$\le \left(2\,\widetilde{\omega}(\alpha) + \frac{\widetilde{\omega}(R)}{1 + p}\right)\|x_1 - x_0\|.$$

Moreover, since

$$\|I - \widetilde{\Gamma}F'(x_1)\| \le \|\widetilde{\Gamma}\|\,\|F'(\widetilde{x}) - F'(x_1)\|$$

$$\le \frac{\beta}{1 - \beta\widetilde{\omega}(\alpha)}\,\widetilde{\omega}(\|x_1 - \widetilde{x}\|)$$

$$\le \frac{\beta}{1 - \beta\widetilde{\omega}(\alpha)}\,(\widetilde{\omega}(\alpha) + \widetilde{\omega}(R))$$

$$< 1,$$

the Banach lemma on invertible operators ensures that Γ_1 exists and

$$\|\Gamma_1\| \leq \frac{\beta}{1 - \beta(2\,\widetilde{\omega}(\alpha) + \widetilde{\omega}(R))}.$$

Furthermore, from (8.12), we obtain

$$\|x_2 - x_1\| \leq \|\Gamma_1\|\|F(x_1)\| \leq \widetilde{P}\|x_1 - x_0\|,$$

$$\|x_2 - x_0\| \leq \|x_2 - x_1\| + \|x_1 - x_0\| \leq (1 + \widetilde{P})\|x_1 - x_0\| \leq (1 + \widetilde{P})\eta < R,$$

where $\widetilde{P} = \dfrac{\beta\left(2\,\widetilde{\omega}(\alpha) + \frac{\widetilde{\omega}(R)}{1+p}\right)}{1 - \beta(2\,\widetilde{\omega}(\alpha) + \widetilde{\omega}(R))}$. Thus, $x_2 \in B(x_0, R)$.

The remaining part of the proof is similar to that of Theorem 8.3 where we replace Q by $\widetilde{Q} = \dfrac{2\beta(\widetilde{\omega}(\alpha) + \widetilde{\omega}(R))}{1 - \beta(2\,\widetilde{\omega}(\alpha) + \widetilde{\omega}(R))}$. ∎

Bibliography

1. F. Aguiló and A. Miralles, *Consideraciones geométricas acerca del método de Newton* (Spanish), Gac. R. Soc. Mat. Esp., 7, 1 (2004) 247–260.
2. S. Amat and S. Busquier, *Third-order iterative methods under Kantorovich conditions*, J. Math. Anal. Appl., 336, 1 (2007) 243–261.
3. S. Amat, S. Busquier and J. M. Gutiérrez, *Third-order iterative methods with applications to Hammerstein equations: a unified approach*, J. Comput. Appl. Math., 235, 9 (2011) 2936–2943.
4. A. A. Andronow and C. E. Chaikin, *Theory of Oscillations*, Princenton University Press, New Jersey, 1949.
5. I. K. Argyros, *Quadratic equations and applications to Chandrasekhar's and related equations*, Bull. Austral. Math. Soc., 32 (1985) 275–292.
6. I. K. Argyros, *On a class of nonlinear integral equations arising in neutron transport*, Aequationes Math., 36 (1988) 99–111.
7. I. K. Argyros, *The Newton–Kantorovich method under mild differentiability conditions and the Pták error estimates*, Monatsh. Math., 101 (1990) 175–193.
8. I. K. Argyros, *Remarks on the convergence of Newton's method under Hölder continuity conditions*, Tamkang J. Math., 23, 4 (1992) 269–277.
9. I. K. Argyros, *A fixed point theorem for perturbed Newton-like methods on Banach space and applications to the solution of nonlinear integral equations appearing in radiative transfer*, Commun. Appl. Anal., 4, 3 (2000) 297–303.
10. I. K. Argyros, *An improved error analysis for Newton-like methods under generalized conditions*, J. Comput. Appl. Math., 157 (2003) 169–185.
11. I. K. Argyros, *On a theorem of L. V. Kantorovich concerning Newton's method*, J. Comput. Appl. Math., 155 (2003) 223–230.
12. I. K. Argyros, *On the Newton–Kantorovich hypothesis for solving equations*, J. Comput. Appl. Math., 169, 2 (2004) 315–332.
13. I. K. Argyros, *A semilocal convergence analysis for directional Newton methods*, Math Comput., 80 (2011) 327–343.
14. I. K. Argyros and S. Hilout, *Weaker conditions for the convergence of Newton's method*, J. Complexity, 28 (2012) 364–387.
15. I. K. Argyros and S. Hilout, *On the quadratic convergence of Newton's method under center-Lipschitz but not necessarily Lipschitz hypotheses*, Math. Slovaca, 63 (2013) 621–638.
16. I. K. Argyros and D. González, *Unified majorizing sequences for Traub-type multipoint iterative procedures*, Numer. Algorithms, 64, 3 (2013) 549–565.

© The Editor(s) (if applicable) and The Author(s), under exclusive licence to Springer Nature Switzerland AG 2020

J. A. Ezquerro Fernández, M. Á. Hernández Verón, *Mild Differentiability Conditions for Newton's Method in Banach Spaces*, Frontiers in Mathematics, https://doi.org/10.1007/978-3-030-48702-7

17. I. K. Argyros and D. González, *Extending the applicability of Newton's method for k-Fréchet differentiable operators in Banach spaces*, Appl. Math. Comput., 234 (2014) 167–178.
18. S. Banach, *Théorie des Opérations Linéaires*, Monografie Matematyczne, Varsovia, 1932.
19. J. L. Chabert et al., *A History of Algorithms: From the Pebble to the Microchip*, Springer-Verlag, Berlin-Heidelberg, 1999.
20. S. Chandrasekhar, *Radiative Transfer*, Dover Publications, New York, 1960.
21. F. Cianciaruso and E. De Pascale, *Newton–Kantorovich approximations when the derivative is Hölderian: old and new results*, Numer. Funct. Anal. Optim., 24, 7–8 (2003) 713–723.
22. F. Cianciaruso and E. De Pascale, *Estimates of majorizing sequences in the Newton–Kantorovich method*, Numer. Funct. Anal. Optim., 27, 5–6 (2006) 529–538.
23. F. Cianciaruso and E. De Pascale, *Estimates of majorizing sequences in the Newton–Kantorovich method: a further improvement*, J. Math. Anal. Appl., 322 (2006) 329–335.
24. H. T. Davis, *Introduction to Nonlinear Differential and Integral Equations*, Dover Publications, New York, 1962.
25. J. E. Dennis and R. B. Schnabel, *Numerical Methods for Unconstrained Optimization and Nonlinear Equations*, SIAM, Philadelphia, 1996.
26. J. A. Ezquerro and M. A. Hernández, *Recurrence relations for Chebyshev-type methods*, Appl. Math. Optim., 41 (2000) 227–236.
27. J. A. Ezquerro, M. A. Hernández and M. A. Salanova, *Recurrence relations for the midpoint method*, Tamkang J. Math., 31 (2000) 33–41.
28. J. A. Ezquerro and M. A. Hernández, *Generalized differentiability conditions for Newton's method*, IMA J. Numer. Anal., 22, 2 (2002) 187–205.
29. J. A. Ezquerro and M. A. Hernández, *On the R-order of convergence of Newton's method under mild differentiability conditions*, J. Comput. Appl. Math., 197, 1 (2006) 53–61.
30. J. A. Ezquerro, J. M. Gutiérrez, M. A. Hernández, N. Romero, M. J. Rubio, *The Newton method: from Newton to Kantorovich* (Spanish), Gac. R. Soc. Mat. Esp., 13, 1 (2010) 53–76.
31. J. A. Ezquerro, D. González and M. A. Hernández, *A general semilocal convergence result for Newton's method under centered conditions for the second derivative*, ESAIM-Math. Model. Numer. Anal., 47, 1 (2013) 149–167.
32. J. A. Ezquerro, M. A. Hernández-Verón and A. I. Velasco, *An analysis of the semilocal convergence for secant-like methods*, Appl. Math. Comp., 266 (2015) 883–892.
33. J. A. Ezquerro and M. A. Hernández-Verón, *On the accessibility of Newton's method under a Hölder condition on the first derivative*, Algorithms, 8, 3 (2015) 514–528.
34. J. A. Ezquerro and M. A. Hernández-Verón, *How to improve the domain of parameters for Newton's method*, Appl. Math. Lett., 48 (2015) 91–101.
35. J. A. Ezquerro and M. A. Hernández-Verón, *Enlarging the domain of starting points for Newton's method under center conditions on the first Fréchet-derivative*, J. Complexity, 33 (2016) 89–106.
36. J. A. Ezquerro and M. A. Hernández-Verón, *On the domain of starting points of Newton's method under center Lipschitz conditions*, Mediterr. J. Math., 13 (2016) 2287–2300.
37. J. A. Ezquerro and M. A. Hernández-Verón, *Newton's Method: An Updated Approach of Kantorovich's Theory*, Frontiers in Mathematics, Birkhäuser/Springer, Cham, 2017.
38. J. A. Ezquerro and M. A. Hernández-Verón, *A study of the influence of center conditions on the domain of parameters of Newton's method by using recurrence relations*, Adv. Comput. Math., 43, 5, (2017) 1103–1129.
39. I. Fenyö, *Über die Lösung der im Banachschen Raume definierten nichtlinearen Gleichungen*, Acta Math. Hungar., 5 (1954) 85–93.
40. Gallica-Math: OEuvres Complètes, Biblioteca Numérica Gallica de la Bibliothèque Nationale de France, http://mathdoc.emath.fr/OEUVRES/

41. M. Ganesh and M. C. Joshi, *Numerical solvability of Hammerstein integral equations of mixed type*, IMA J. Numer. Anal. 11 (1991) 21–31.
42. D. Greenspan, *Introductory Numerical Analysis of Elliptic Boundary Value Problems*, Harper and Row, New York, 1965.
43. J. M. Gutiérrez and M. A. Hernández, *Newton's method under weak Kantorovich conditions*, IMA J. Numer. Anal., 20 (2000) 521–532.
44. J. M. Gutiérrez and M. A. Hernández, *Newton's method under different Lipschitz conditions*, in: Numerical Analysis and its Applications, Lectures Notes in Comput. Sci., 1988 (2000) 368–376.
45. J. M. Gutiérrez, Á. A. Magreñán and N. Romero, *On the semilocal convergence of Newton–Kantorovich method under center-Lipschitz conditions*, Appl. Math. Comput., 221 (2013) 79–88.
46. M. A. Hernández, *Relaxing convergence conditions for Newton's method*, J. Math. Anal. Appl., 249 (2000) 463–475.
47. M. A. Hernández, M. J. Rubio and J. A. Ezquerro, *Secant-like methods for solving nonlinear integral equations of the Hammerstein type*, J. Comput. Appl. Math., 115, 1–2 (2000) 245–254.
48. M. A. Hernández, *The Newton method for operators with Hölder continuous first derivative*, J. Optim. Theory Appl., 109 (2001) 631–648.
49. L. V. Kantorovich, *On Newton's method for functional equations*, Dokl. Akad. Nauk. SSSR, 59 (1948) 1237–1240.
50. L. V. Kantorovich, *Functional analysis and applied mathematics* (Russian), Uspekhi Mat. Nauk, 3 (1948) 89–185.
51. L. V. Kantorovich, *On Newton's method* (Russian), Trudy Math. Inst. Steklov, 28 (1949) 104–144.
52. L. V. Kantorovich, *The majorant principle and Newton's method* (Russian), Dokl. Akad. Nauk. SSSR, 76 (1951) 17–20.
53. L. V. Kantorovich, *Some further applications of principle of majorants* (Russian), Dokl. Akad. Nauk SSSR, 80 (1951) 849–852.
54. L. V. Kantorovich, *On approximate solution of functional equations* (Russian), Uspekhi Mat. Nauk, 11 (1956) 99–116.
55. L. V. Kantorovich, *Some further applications of Newton's method* (Russian), Vest. LGU, Ser. Math. Mech., 7 (1957) 68–103.
56. L. V. Kantorovich and G. P. Akilov, *Functional Analysis in Normed Spaces* (Russian), Fizmatgiz, Moscow, 1959; translated by D. E. Brown and A. P. Robertson, Pergamon Press, Oxford, 1964.
57. L. V. Kantorovich and G. P. Akilov, *Functional Analysis*, Pergamon Press, New York, 1982.
58. H. B. Keller, *Numerical Methods for Two-point Boundary Value Problems*, Dover Publications, New York, 1992.
59. J. H. Mathews, *Bibliography for Newton's method*, http://mathfaculty.fullerton.edu/mathews/n2003/newtonsmethod/Newton'sMethodBib/Links/Newton'sMethodBib_lnk_2.html
60. J. M. McNamee, *A bibliography on roots of polynomials: Newton's method*, http://www1.elsevier.com/homepage/sac/cam/mcnamee/02.htm
61. O. Neugebauer and A. Sachs, *Mathematical Cuneiform Texts*, American Oriental Society, New Haven, Connecticut, 1945.
62. J. M. Ortega, *The Newton–Kantorovich theorem*, Amer. Math. Monthly, 75 (1968) 658–660.
63. A. Ostrowski, *Über die Konvergenz und die Abrundungsfestigkeit des Newtonschen Verfahrens*, Rec. Math., 2 (1937) 1073–1095.
64. A. Ostrowski, *Über einen Fall der Konvergenz des Newtonschen Näherungsverfahrens*, Rec. Math., 3 (1938), 254–258.
65. B. T. Polyak, *Newton–Kantorovich method and its global convergence*, J. Math. Sciences, 133, 4 (2006) 1513–1523.

66. A. D. Polyanin and A. V. Manzhirov, *Handbook of Integral Equations*, CRC Press, Boca Raton, 1998.
67. D. Porter and D. Stirling, *Integral Equations: A Practical Treatment, from Spectral Theory to Applications*, Cambridge University Press, Cambridge, 1990.
68. F. A. Potra and V. Pták, *Nondiscrete Induction and Iterative Methods*, Pitman Publishing Limited, London, 1984.
69. J. Rashidinia and M. Zarebnia, *New approach for numerical solution of Hammerstein integral equations*, Appl. Math. Comput., 185 (2007) 147–154.
70. L. B. Rall, *Computational Solution of Nonlinear Operator Equations*, Robert E. Krieger Publishing Company, Michigan, 1979.
71. W. C. Rheinboldt, *A unified convergence theory for a class of iterative processes*, SIAM J. Numer. Anal., 5 (1968) 42–63.
72. J. Rokne, *Newton's method under mild differentiability conditions with error analysis*, Numer. Math. 18 (1972) 401–412.
73. E. Schröder, *Über unendlich viele Algorithmen zur Auflösung der Gleichungen*, Math. Ann. 2 (1870) 317–365.
74. J. J. Stoker, *Nonlinear Vibrations*, Interscience-Wiley, New York, 1950.
75. J. L. Varona, *Graphic and numerical comparison between iterative methods*, Math. Intelligencer, 24, 1 (2002) 37–46.
76. S. Wolfram, *The Mathematica Book, 5th ed.*, Wolfram Media/Cambridge University Press, 2003.
77. T. Yamamoto, *Historical developments in convergence analysis for Newton's and Newton-like methods*, J. Comput. Appl. Math., 124 (2000) 1–23.
78. L. Yau and A. Ben-Israel, *The Newton and Halley methods for complex roots*, Amer. Math. Monthly, 105, 9 (1998) 806–818.
79. T. J. Ypma, *Historical development of the Newton–Raphson method*, SIAM Review, 37, 4 (1995) 531–551.

Printed in the United States
By Bookmasters